Our Greatest Challenge

Our Greatest Challenge

Eliminate CO$_2$ Emissions

Eduardo Saucedo

Printed in the United States of America

Library of Congress Cataloging-in-Publication Data

Saucedo, Eduardo.

Our greatest challenge: Eliminate CO_2 emissions.

ISBN – 13:978-151 884 3631
ISBN – 10:151 884 3638

To Susan,

My love, wife, confidant and friend, who is my unwavering fan, my staunchest supporter, my admiral and editor in chief, who always allowed me to chase rainbows, even if it meant I was putting our finances in peril.

Our Greatest Challenge

Eliminate CO$_2$ Emissions

Contents

LIST OF FIGURES

LIST OF TABLES

Our Greatest Challenge
Eliminate CO_2 Emissions

Introduction

We are fortunate to be living in the best times humanity has ever experienced. Never before has the human race enjoyed such a bountiful and opulent time, with great advances and developments, enjoying the longest average life span and the highest standard of living ever. In great measure, we have to thank fossil fuels for this.

Unfortunately, burning fossil fuels produces CO_2 which stays in the atmosphere for hundreds of years, where it blocks some of the infrared radiation from leaving the Earth, trapping heat and contributing to global warming. Arrhenius, a Swedish scientist, was the first to postulate the greenhouse effect in 1896; he was happy to envision that Sweden might be warmed.

For eons, the temperature has been kept in balance by a natural cycle, with plants absorbing CO_2 and using it to create organic matter. The amount of CO_2 released today (about 35 billion tons per year or 35 GTY) has overwhelmed the natural cycle, and CO_2 is slowly, but inexorably, creeping up from the steady concentration of the last ten thousand years of about 280 parts per million ("ppm") to 400 ppm in 2014. As predicted, the average temperature of the Earth has slowly increased, with the ten warmest years ever recorded occurring in this century. The only way to ameliorate global warming (climate change is more politically correct) is to drastically reduce or eliminate future CO_2 emissions. Elimination might sound draconian, but unless we aim for it, we will not be able to solve the problem.

Humanity's greatest challenge, OUR greatest challenge, is how to preserve the standard of living we (all 7.2 billion of us) have gotten accustomed to without burning fossil fuels.

Notwithstanding past great achievements (the Pyramids, the Great Wall of China, the splendors of Athens and Rome, etc.), prior to the industrial revolution, life was rather simple, mostly confined to what could be made, grown and collected locally. Venice's merchants went to Asia, bringing back some camel loads of silk and spices, and a few shiploads of sugar were brought from the Caribbean to England.

With the utilization of mined coal, and then with the use of oil, a new era began. Nowadays, it is possible to find in the US flowers flown in from Colombia, chocolate processed by Swiss companies, asparagus available throughout the year, clothes made in China or the Philippines, glacier water from New Zealand and millions of people visiting exotic places far away from home. None of it would have been possible without access to abundant and cheap energy.

I am an engineer interested in renewable energy. I have studied in depth many possibilities and have concluded that it is a huge problem with no easy solution, and that most renewable energy advocates are many times naïve, misinformed, overly optimistic or dreamers. I want to present a balanced view of actual alternatives, noticing that unfortunately what is available today is not perfect and much remains to be solved or invented. However, rather than stopping there, I want to explore possible costs, time requirements, financing alternatives and implementation stages.

This is not another book on renewable energy. It is not another technical book full of graphs and equations, although you might find some here. It is not another alarmist or catastrophe-minded book proclaiming that the end of the oil days is upon us we better start buying candles to see at night and horses for transportation. It is not another positive-thinking book suggesting that geologists meditate to find more oil. Finally, it is not an environmentalist book simply advocating a magical hydrogen economy.

It is a book that will discuss the limitations of the available alternatives to provide the energy needed to meet humanity's future needs, recognizing that it is a very complex problem and that the existence of easily extractable, transportable and convertible energy has had a tremendous impact on the explosive growth in the number of people, on their wealth, comfort, health and possibilities that surpass what most people in the nineteenth century could have imagined. It is impossible to duplicate the versatility and energy density of fossil fuels, but it does not mean that we have to do without; rather, we have to adapt to a new reality and sacrifice now so that we all can enjoy a brighter future.

In this book, I have concentrated the discussion on CO_2, but it is not the only green house gas that we need to be concerned about. As a matter of fact, water vapor is the worst offender, while methane and nitrous oxide are 23 and 296 times worse than CO_2, respectively. Fortunately, they stay in the atmosphere only a fraction of the time. Water vapor leaves as rain, methane burns to produce CO_2 and water and nitrous oxide gets washed out with rain in the form on nitric acid. However, the changes needed to become a carbonless society will also ameliorate the emission of both methane and nitrous oxide.

My hope is that this book can serve as the basis to start real, serious, positive discussions. The challenge is huge. I believe I have not exaggerated even a tiny bit by stating that this is the greatest challenge humanity has ever faced. The cost is going to be staggering; the shuffling of the economy will cause severe pain and suffering to many, but also provide new opportunities, and the entire world has to cooperate. CO_2, whether it is released in Argentina, China, Europe, the US or Zambia, goes into the atmosphere and affects us all. Notwithstanding military or economic power, no country can be exempted. This might be the beginning of a new world order, of a new "we are all in it together" way of thinking that might integrate the world and bring huge benefits to smaller communities, create local jobs and promote consumption of locally made products and produce.

This book concludes that there are alternatives. We do not have to give up heating (or cooling) our houses. When needed, there will be ways of transporting us and goods across the country. We could maintain a similar standard of living as that to which we are accustomed. We will have to pay the price for it, but we can afford it, and by making reasonable choices and sacrifices, we can improve our world.

This book assumes that we will eventually rise to the challenge and decide to eliminate the emission of CO_2 and look at possible mechanisms to achieve that goal, and therefore will explore in depth the three proposed mechanisms: cap and trade, carbon tax and carbon rationing.

There was no mistake in the previous paragraph. I purposely and clearly stated "ELIMINATE" and did not try to sweeten the bitter pill by saying "reduce," "ameliorate" or even just "arrest" CO_2 emissions. An indebted family or a country cannot reduce their debt by borrowing less. It has to stop borrowing and start paying. If the CO_2 stays in the atmosphere for hundreds of years, what is the point of making some sacrifices today if the end result is that the world is doomed in 2050, 2100 or 2150? I do agree there is a good possibility that we will eventually learn how to harness fusion energy in the next one or two hundred years, but the question is whether the human race will survive that long.

I am aware that none of the current available technologies is capable of substituting for the reliability of fossil fuels and that electricity storage is not feasible at the level needed. However, I will be proposing that we use wind, solar, thermal and nuclear power and combined with hydrogen production by electrolysis. The production of hydrogen would be price sensitive, reacting to price changes, absorbing surplus energy when available and not consuming electricity when the price passes the chosen threshold.

Organization

The energy topic is complex and we take for granted its presence. Even before we leave home in the morning, energy is there to help us: it wakes us up with an electric alarm clock, provides a hot shower, makes the coffee and toasts the bread and show the local news on TV while we have breakfast. Any change to the current energy policies is going to affect us one way or the other.

Chapter One talks about the side effects of burning fossil fuels, namely the irrefutable increase of CO_2 concentration in the atmosphere; presents data showing that foreseen consequences are manifesting and shows historical data of CO_2 emissions by countries and per capita for selected countries.

Chapter Two considers the complicated multi-faceted problem of reducing/eliminating CO_2 emissions. It has a technical component because renewable energy is less versatile and/or reliable than fossil fuels; an economic component because the needed increase in the price of energy would affect everything; a financial component because a really, really enormous investment will have to be made; and also a human component, because we need our leaders to start leading us rather than pretending that everything is going to be fine.

Chapter Three covers energy in general terms, consumption per capita and as a fraction of the GDP; its influence on the price of most commodities; its inelasticity; presents a summary of energy in the US consumption and reserves, and; per capita consumption as primary energy, which is translated into such needs as heating, electricity, transportation and other uses.

Chapter Four talks about the fossil fuel era, about oil reserves, natural gas and coal and concludes that while reserves seem large, they are concentrated in a few countries, and given the possibility that production might decline, strongly recommends that we start studying other alternatives.

Chapter Five discusses electricity, which is not a primary source of energy, but the most adaptable carrier of energy, giving some details about its structure, its evolution and the variability of demand during the day and by season, and the lack of inexpensive or practical ways to store electricity or energy.

Chapter Six talks goes over solar radiation, a huge source of energy, yet extremely diluted and variable, even without taking into consideration cloud cover.

Chapter Seven analyzes the available forms of renewable energy and discusses their advantages, potential and limitations with respect to replacing fossil fuels.

Chapter Eight talks about the cost of generating, with renewable energy sources, all the electricity we need to replace existing fossil fuel generation but also to produce the hydrogen that will be used for transportation. Finally, it reflects that the astronomical cost the conversion entails cannot be financed by the private sector and would have to be financed by the government and by all of us in the form of a carbon fee.

Chapter Nine discusses the possible mechanisms to reduce the emission of CO2 but also to secure the funding needed to finance the required investment to transform us into a carbonless society.

Chapter Ten expands on the proposed fair and market-driven carbon rationing mechanism and how it can be used to finance the shift to a carbonless society.

Chapter Eleven discusses how the implementation of the carbon rationing scheme worldwide would form the basis for international cooperation by using the rationing mechanism to transfer resources to less developed countries.

Finally, Chapter Twelve presents conclusions and urges everyone to rise to the challenge for the benefit of future generations.

Units and Calculations

For simplicity I like rounding things up and down to have simple numbers. The order-of-magnitude calculations do not change much. The end result, whether it is expressed as carbon foot print in kg CO_2 per day per person can be 60.63 kg/p d or 60, is about the same. The population of the U.S. can be presented as 322,583,006 as of July 1, 2014 or 320 million. The end result might be off 1% or 5%. When we talk about $1.7 trillion, it is a huge amount, but whether it is really $1.684 or $1.721 trillion, it does not much change the fact that it is a large number representing about 10% of the GDP of the U.S. Providing more digits implies more precise numbers that are not available. The installed power generation equipment listed by the Energy Information Agency is 986,000 MW (one million MW is about the same) and generates 4.1 PWhr/y (the exact number is 4.065 PWhr/y), which, expressed as average utilization per year, yields 47.03% when using exact numbers (the year has 365.25 days) or 46.8% if using round numbers. I will be using 47%. Some of the calculations eventually will have to be translated into a suggested CO_2 price. Whether the number is 14, 15 or 16¢/kg makes no difference. When translated into how much we will have to pay, it comes to a good chunk of money. Others with better access to statistics can refine the numbers.

The scientific unit of energy or work is a Joule. Power is a measure of work per unit time. A system capable of doing twice as much work every second needs twice the amount of power. The unit of power is a joule/sec or Watt. Ten kW is the unit of power needed to lift one hundred kg of mass, subject to the force of gravity, one meter in one second. One kW is a unit of power, but one kWh is a unit of energy. One kW is 1,000 watts or 1000 joules/sec of power, which when multiplied by 3,600 seconds per hour, gives you 3.6 MJ as energy.

We use MW to indicate how big a system is, but it does not mean the power is available all the time. A one MW thermoelectric plant could generate some 8,000 MWh per year, but a one MW wind turbine that can generate that power when the wind is blowing at a speed of 10 m/s will not be able to generate any electricity on a calm day. A one MW wind turbine usually generates about 2,200 MWh per year. PV panels are usually rated in peak watts (the output produced when the irradiation is 1000 w/m^2 h measured at standard conditions), but can only generate between 1,000 to 1,500 watts per year per peak watts. One base load thermoelectric or nuclear plant can usually generate about 8,000 hours per year, while a peaking plant might only work a few hours a day on some very hot days in the summer.

Average energy consumption is usually misleading. The average consumption in an upper middle class house might be 3 kWh or about 26.3 MWh/year, but the instantaneous power needed by that household when the air conditioning is working, the family is using an electric oven, microwave, the TV is on, clothes are being washed while others are being dried and the water heater is replenishing the hot water could sometimes hit 20 kW.

Energy can be expressed in a multitude of ways, from gallons of gasoline, quadrillion BTUs or pentawatt hours, and big numbers have different meanings in the world. All the units can be converted back and forth. For simplicity and convenience (I am writing in the US), I will adhere to the American convention with respect to large numbers – so one billion is 1,000 million and one trillion is 1,000 billion or 10^{12}. For the purpose of energy and big numbers, I will try to use as much as possible tons of oil equivalent or TOE (or toe) and when translated into smaller units as per capita per year, I will use gallons of oil equivalent/person year or simply g/p d (I used 0.9 as oil's density). For electricity, I will use the SI prefix convention, with peta, tera, giga, mega and kilo meaning 10^{15}, 10^{12}, 10^9, 10^6 and 10^3, respectively.

Acknowledgements

I want to thank the professors at Duke that encourage me to write this book and John Lee for his dedication, sharp eyes and diligence in reading and editing the manuscript. Any errors still there are mine.

Chapter 1 CO_2 in the Atmosphere

We live in the most prosperous era ever experienced by mankind. The standard of living has never been greater for many, with bright and warm houses at night, fresh food from all corners of the world, wondrous devices that can make a cup of coffee in seconds, being able to instantly communicate or visit with relatives even if they are on the other side of the world, with a large portion of the population living well past retirement age. All of it has been possible thanks to the harnessing of fossil fuels.

The ability to control fire set us apart. It allowed us to extract more nutrition from gathered foods and reduced the time needed for food gathering and/or digesting, reduced our dependence of migrating back and forth following herds and likely contributed to our development. Every time fuel is burned, it releases some CO_2. When there were fewer humans burning wood, plants maintained the balance by removing CO_2 from the atmosphere and fixing it in organic matter. As the population grew and the need of fuels for other purposes increased, humanity overwhelmed the capacity of the natural cycle to absorb emitted CO_2, which has slowly but steadily increased its concentration in the atmosphere.

Since Arrhenius, scientists have refined the calculations and confirmed the results with other methods (for example CO_2 separated from deep ice rings), and today the vast majority of scientists in the field agree that the increased CO_2 concentration will trap more heat in the Earth's atmosphere.

CO_2 is not the only gas in the atmosphere trapping heat. Methane, nitrous oxide and water vapor can trap much more, but water vapor fluctuates continuously and usually comes down as rain, nitrous oxide eventually forms nitric acid that is brought down with the rain and methane usually burns over time and only lasts about a decade. CO_2, on the other hand, might remain in the atmosphere for centuries. Methane is produced by leaks in gas transmissions, by cattle and waste (among other things); and nitrous oxide is produced in the high temperature boilers in thermoelectric plants or internal combustion engines.

We, humanity, face our biggest challenge ever. Continuing on the present trajectory will further aggravate the consequences and yet, even if there were the political will to give up the advantages that fossil fuels provide and somehow voluntarily cutting back the CO_2 emissions to medieval rates, the pain, the economic contraction and the suffering that it would entail makes it quite difficult to have the courage to implement this monumental change. The best we can hope is for a reasonable transition period.

CO₂ Concentration

Early measurements of CO_2 concentration showed variations between locations. Some variations were attributed to wind, temperature, the time of the day or even surrounding vegetation or proximity to cities or industrial development. With better instrumentation and sampling at different places, dates and times during the day, Keeling detected even a strong diurnal variation. After concluding that a remote site was needed, Keeling proposed to measure CO_2 in Hawaii and Antarctica. The Mauna Loa station has continuously monitored CO_2 concentrations since 1958. Data from 1974 is shown in Figure No. 1.

The figure shows a yearly fluctuation, attributable to the large intake of CO_2 during the growing season in the northern hemisphere, which has a larger land mass, and consequently more biomass, than the southern hemisphere. Concentration of CO_2 in the atmosphere peaks in May and starts decreasing until October due to the fixing of CO_2 by photosynthesis.

CO_2 concentration of in the atmosphere has shot up, starting with the Industrial Revolution and has reached 404 ppm in June 2015.

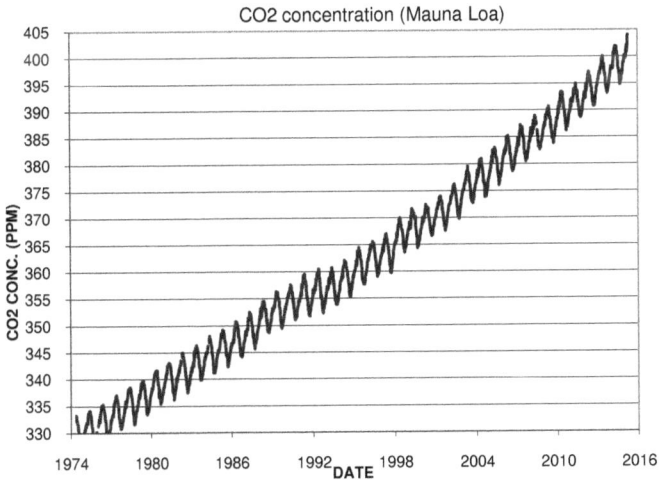

Figure 1 CO$_2$ Concentration

From CO$_2$ entrapped in ice, scientists have been able to gather evidence of the CO$_2$ concentration in the atmosphere over the previous 800,000 years. It has oscillated regularly from about 180 parts per million (ppm) during deep glaciations periods to 280 ppm during interglacial periods as shown in Figure No. 2.

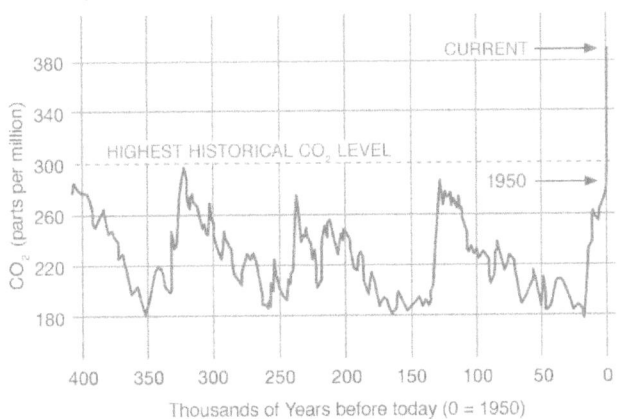

Figure 2 Historical Long Term CO$_2$ Concentration

The World Bank compiled per capita emission of CO$_2$ by country since 1960. The average emission per capita has increased from about 3 tons of CO_2 per person per year (3 tons/p y) in 1960 to 5 tons/p y in 2010. The world population in 1960 was 2.982 billion and represented global emissions of about 9 GTY (9 billion tons per year). In 2010, the world population was roughly 7 billion and global emissions were 34 GTY.

The emissions are not homogeneous. Figure No. 3 below shows the emissions per capita for selected countries from 1960 to 2010.

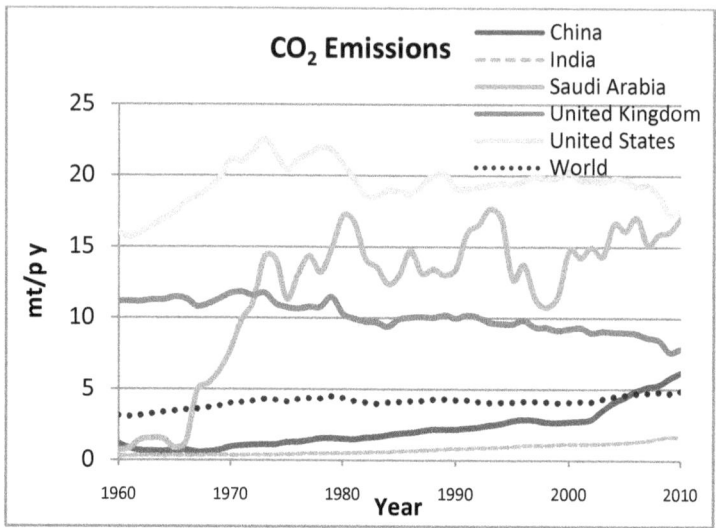

Figure 3 CO_2 Emissions per Capita (metric t/p y)

Developed countries have larger emissions per capita than poorer countries. Their industry is more developed, their income per capita is larger, and they can afford more cars and heating and cooling of larger houses. Oil producing countries are large emitters per capita, with Qatar being the largest. It requires energy to produce energy. Excluding a few oil producing countries with small populations, the US was by far the largest emitter per capita, with Canada and Australia following rather close behind. During the 1990s, the US emissions represented more than 25% of global emissions. With its drastic growth, China is now the largest emitter, overtaking the US in 2012, and the latter's share has shrunk to about 20% of global emissions. By contrast, many

African nations emit less that 1 ton/p y, and the poorest countries only a fraction of 1 ton/person year.

From the global perspective, Table No. 1 below shows data compiled by the National Oceanic and Atmospheric Administration (NOAA) for CO$_2$ in the atmosphere, the World Bank for emissions per capita and the United Nations for population

Global CO2 Data							
Year	Atmospheric CO2 (ppm)	Average Growth Rate	CO2 Per Capita (tons)	Average Growth Rate	Population (million)	Average Growth Rate	CO2 Emited (Gty)
1974	329.69	1.0050	4.2215	1.0232	3,995	1.0186	16.865
1984	344.41	1.0044	4.0386	0.9956	4,776	1.0180	19.288
1994	358.82	1.0041	4.0463	1.0002	5,661	1.0171	22.906
2004	377.52	1.0051	4.4646	1.0099	6,435	1.0129	28.730
2014	398.58	1.0054	4.8929	1.0092	7,283	1.0125	35.635
Cumm.		1.0048		1.0037		1.0151	1.0189

Table 1 CO$_2$ Emitted per Year

While the population grew at an average 1.5% annually during the last 40 years and the per capita emission by 0.37% per year, the CO$_2$ emitted grew by 1.88% but the CO$_2$ atmospheric concentration by only 0.47%. The difference is easily explained by the size of the atmosphere. Two trends are worrisome: (i) the amount of CO$_2$ emitted has more than doubled in the last 40 years, and; (ii) the rate of increase in concentration in the atmosphere has been increasing steadily.

Irrefutable Evidence

The previous paragraphs presented some irrefutable evidence that: (i) the atmospheric concentration of CO$_2$ has been growing steadily since the onset of the Industrial Revolution; (ii) we might be entering into unchartered territory - the planet has not experienced such a high level of CO$_2$ in the past 800,000 years; (iii) mankind has been emitting larger and larger amounts of CO$_2$ every year (on a cumulative basis, we have emitted about 1,000 GT of CO$_2$ into the atmosphere in the last 40 years), and; (iv) several of the predicted consequences of higher concentration of CO$_2$ in the atmosphere are manifesting.

Temperature Change

The graph below shows the change of temperature anomaly from 1880 to date. Positive change denotes higher temperature (anomaly is more politically correct than global warming).

Figure No. 4 shows that the anomaly[1] (i.e. the change over the mean or average) has been growing steadily since 1910. Skeptics have been quick to point out that the anomaly did not continue increasing in the second half of the last decade, promptly concluding that it had paused or reached a hiatus, that it was only a typical weather temperature variation and that it was slowing down, going from 0.114°C/decade to 0.104°C/decade, and quickly dismissing the fact that eight of the last ten warmest years in recorded history have occurred in this decade.

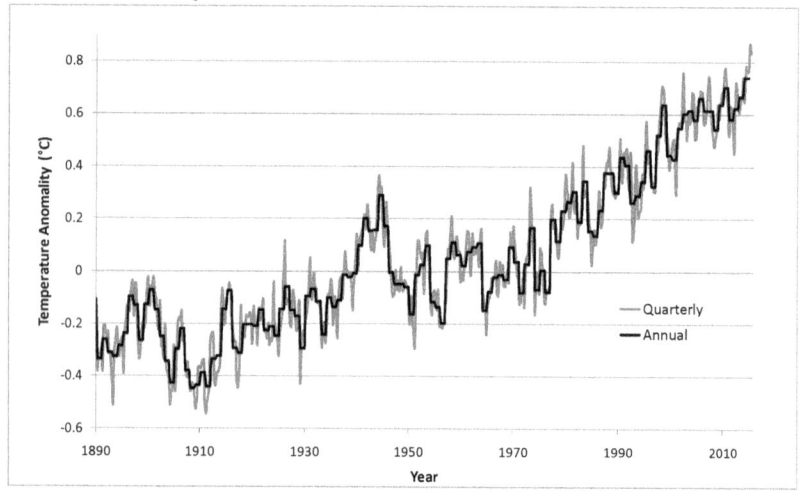

Figure 4 Temperature Change Anomaly

Ice Extent

Figure No. 5 shows the variation of the Arctic ice extent (the ocean surface covered by ice) from 1979 to date, as provided by NASA[2]. The graph shows unequivocally that the Arctic ice extent is shrinking. The possibility of transiting through the North

---

[1] NOAA's National Center for Environmental Information
[2] NASA Cryosphere Science Research Portal

West Passage is real. Several boats traversed it in the early 2000s, and a commercial vessel made it through in 2013. The minimum ice coverage was reported in September 2012 at only 3.39 million km^2 (less than half the size of the US). The skeptics were quick to point out that, by contrast, the Antarctic ice extent has slightly expanded and reaching a peak at its maximum that is about 1 million km^2 larger than the 1981-2010 average, claiming simplistically that one cancels out the other. It is not that simple. The Antarctic ice changes more during the year; the ice sits on the landmass rather than floating on a sea surrounded by land, and even the thickness of the ice is different.

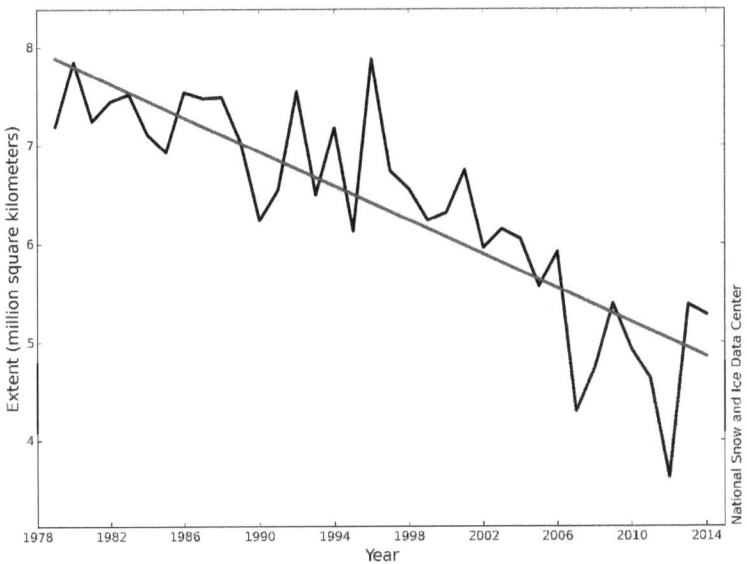

Figure 5 Arctic Ice Extent

Greenland Ice Melt

It appears than Greenland's ice mass is melting faster than in previous periods. The glaciers are noticeably moving faster and many new large lakes form in the summer within the ice mass. The most comprehensive data has been provided by the Grace satellite (Gravity Recovery and Climate Experiment), detecting variations in the strength of the force of gravity in Greenland.

According to the measurements in the Arctic Report Card[3], Greenland has lost some 3,000 GT of ice in the period 2002-2014, as seen in Figure No. 6.

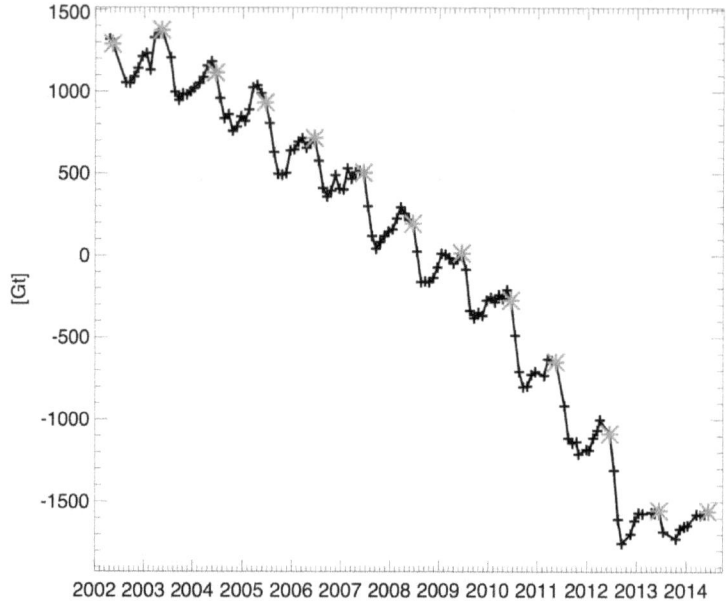

Figure 6 Greenland Ice Melt

Sea Level

Another prediction of the consequences of global warming is a rise in sea level. There are two reasons: (i) as the temperature increases, water expands, and; (ii) the melted ice sheets in Greenland and Antarctic reaching the sea increase the volume of water. Figure No. 7 shows that the sea level is indeed rising, having risen about 20 cm (eight inches) in the last century.

The skeptics would quickly point out that we are making a big deal of a rise of about 0.2cm/y, when the tide alone rises twice a day a meter or more and happily conclude that the sea rise does not appear to be that worrisome!

---

[3] Tedesco et al (Arctic Report Card 2014)

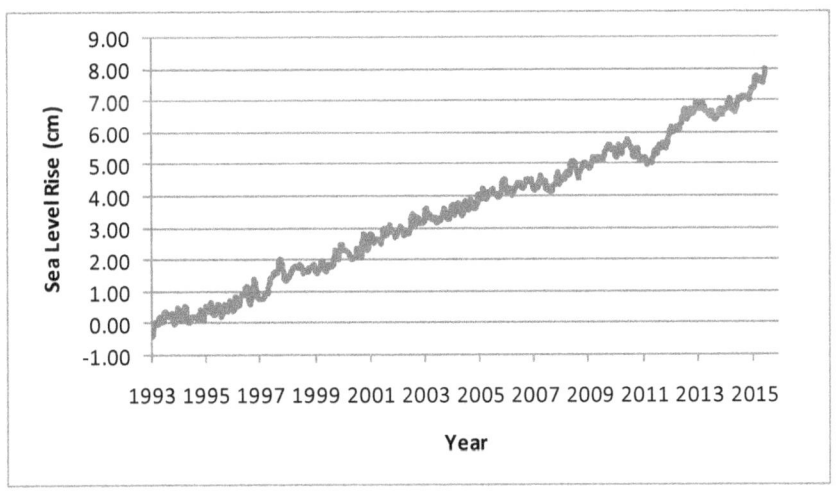

Figure 7 Sea Level Rise

Consequences

Scientists are normally conservative fellows who carefully choose their words. Several have postulated some dire consequences if the concentration of CO$_2$ in the atmosphere continues its unabated rise. They have published their findings in respected scientific publications and stand behind their findings, though other scientists might dispute the conclusions, methodologies, assumptions or parameters used to arrive at those conclusions.

The scientific method allows anybody to postulate theories about any subject, but for the theory to be accepted by the scientific community, it must withstand the scrutiny of experts in the field. The heliocentric theory is a good example of the application of the scientific method. For centuries, the geocentric theory (the Earth is the center of the universe) was the only accepted theory. Nicolaus Copernicus first postulated the heliocentric theory (that the Sun is the center) in 1543, and Galileo Galilei's treatise (1642), finally published at the end of his life, showed phases in Venus' transit (proving that Venus moves around the Sun, not the Earth) and that there were moons circling Jupiter, proving that

17

the Earth is not the center of everything. Many attempts to discredit or provide another explanation failed, and eventually the scientific consensus became that the Earth rotates around the Sun.

There is a plethora of books that talk about the awful things that could happen in the future if global warming remains uncontrolled. Some of those books are really scary and should be best kept in the library's horror section! The Earth is getting warmer; yields of crops are going to fall; there will be more fires; the Amazon forest is going to suffer droughts; there will insufficient snow melt to irrigate crops; the rising ocean temperature and acidification are hurting fisheries and corals; the sea rise will flood Florida, New York, Venice, Bangladesh and erase several Polynesian islands; people will have to retreat from shores, causing massive numbers of refugees; a lot of the infrastructure is going to be flooded, etc. Other books, even more alarming, are talking about tipping points and positive and negative feedbacks, for example the release of methane trapped in the permafrost or the shutdown or slowdown of the North Atlantic thermohaline circulation, not to mention an increased frequency of extreme events which will increase the risk of flooding, drought, erosion, wildfires or nutrients and pollutants reaching the coast.

Any agricultural school can tell you that there are optimal temperatures to grow wheat or corn, and with higher temperatures, the yield drops; marine biologists can test the hypothesis that acidification of the oceans affects the capability of crustaceans to form shells and that warmer water affects corals; loss of ice is affecting polar bears; the snow cover in California's Sierra mountains, the Himalayas and the Andes is shrinking, and therefore the snow melt has lessened. All of the predictions have been built on a very solid scientific base (what sometimes is questioned are the parameters used in the models), therefore we should be worried.

For honest leaders, the potential catastrophe is paralyzing because there are no quick alternatives. If they were to decree that it is illegal to burn fossil fuels, life as we know it would stop instantaneously. No television or air conditioning or heating, no banks, no running water, within days all shelves in the grocery stores would be bare. As we will show later, even a slow approach, 50% reduction by 2050, is expensive and very complex, and it might even be too late.

Failed responses

Many people have been concerned about climate change for some 30 to 40 years. The United Nations Framework Convention on Climate Change (UNFCCC) was created in 1992 to cooperatively consider what could be done to limit global temperature increases. The efforts culminated in the signing in 1997 of the Kyoto Protocol, placing politically acceptable mild curbs on emissions of greenhouse gases. Specifically, the protocol set legally binding targets for developed countries to reduce greenhouse emissions within seven years to about 5% below 1990 levels. While President Clinton signed the agreement on behalf of the US, the protocol was not ratified by the Senate. President Bush considered the protocol harmful to the US economy and never asked for its ratification in the Senate.

Lack of leadership and recriminations hampered progress, and at the end of the decade, rather than a reduction of CO$_2$ emissions, they had grown by 24%. Canada withdrew its support of the protocol in 2011. Efforts to renew the protocol have failed to reach agreement in subsequent meetings in Copenhagen, Lima and Rio. The positions of the parties have somewhat stiffened, with less developed countries seeking either exemptions or compensation, and developed countries agreeing to some voluntary assistance if all participate. The forthcoming 2015 meeting in Paris has asked countries to pledge goals in advance. Catchy phrases like 50 by 50 (50% by 2050), or 30 by 30 have been heard, many to start in 2020. A few countries, notably Germany and Norway, perhaps more concerned, have made significant strides in greening their generation of electricity.

Political realities are what they are. Developed countries started earlier and emit more per capita than less developed ones. But we are all on the same planet and unless we all work together to eliminate emissions, the planet might not be able to sustain the kind of flora and fauna we are accustomed to, and we might suffer. The discussion has to be about the fastest and cheapest way of eliminating ALL emissions produced by burning fossil fuels for heating, electricity and transportation first, and then for uses in industry and agriculture, hoping that the Earth can still absorb some emissions that would linger longer.

It does not make one bit of difference if the CO_2 is emitted by Albania or Zaire, a developed country or an underdeveloped country, a Northern country or a Southern country, in Asia or Europe. Any CO_2 emitted above the natural capacity of the Earth to recycle is going to stay in the atmosphere for several hundred years, and each molecule will reduce the thermal emission of the planet, therefore contributing to global warming. If the bucket is full, any additional drop overflows. Granted, we cannot stop cold turkey. But there should be urgency in our efforts to avoid triggering events that we will not be able to control and that are likely to cause deaths, pain, damage, losses or hardship to mankind.

The tactic of scaring people by showing possible outcomes for the projections has not worked. It is a long-term threat, and not all of us are wired to think long term. But some fortunately are. There are some who can save money, some who even plan for retirement. Others envisioned building pyramids or cathedrals or the Great Wall of China or Nazca's lines. While the pyramids might have been selfish, medieval cathedrals were long-term projects taking generations to build and the Great Wall diminished long-term future threats. Those visionaries were able to cajole, push or shove aside down-to-earth folks who considered their ideas foolish and a great waste of money and effort for things that they would not see completed. Yet, the planet is full of such wonders, from the Parthenon to the Coliseum, or Saint Paul's or Notre Dame or Chichen Itza or Machu Pichu. More recently, we have witnessed efforts that have changed our world, like the Suez and Panama Canals or allowed us to land on the

Moon and even send probes to Pluto! We have within us the capability to make great efforts or sacrifices for the future. Saving our planet from destruction appears to be a worthy project, lacking only a great leader with the vision, charisma and resources to make it happen.

Urgency

We cannot wait for that visionary leader to be born. We will have to make do with what we have available now because the threat is no longer long term; it is real and imminent. Figure No. 8 below shows the attainable CO$_2$ concentration in the atmosphere as a result of various levels of commitment.

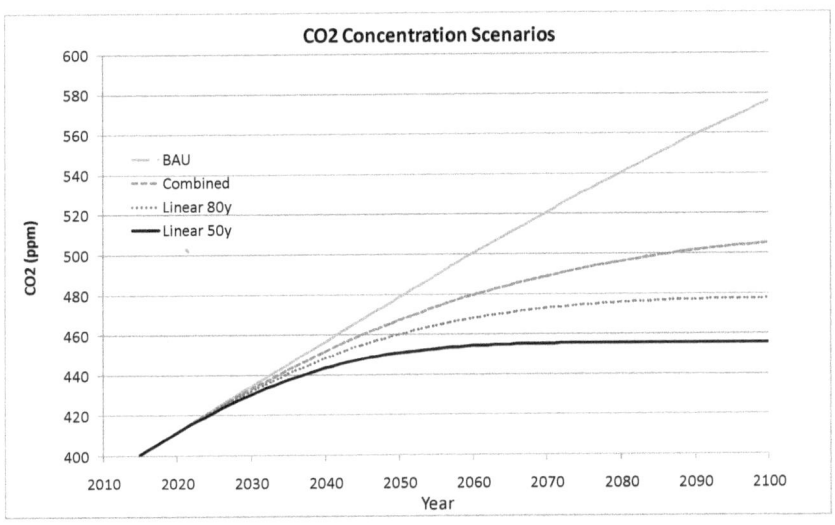

Figure 8 Possible CO$_2$ Concentration Scenarios

The first scenario, labeled Business as Usual ("BAU") is optimistic in its assumptions. It assumes that population growth is going to slowly decrease from about 1.2% annually today to about 0.5% at the end of the century, that the per capita consumption is going to slowly decrease from about 0.5% per year to about 0.35% and that the CO$_2$ concentration growth decreases from about 0.5%

annually today to 0.3%. In this scenario, the CO_2 concentration continues climbing and reaches 576 ppm in the year 2100.

The second line, labeled "Combined," assumes a linear reduction of emissions, starting in 2020, first by the developed world, responsible for 50% of the emissions, reducing half of their emissions, during the first 32 years and the rest in the next 64 years. A second wave of reductions, by less developed countries, responsible for emitting 25% of the CO_2, starts in 2030, following the same reduction pattern as the developed countries. Finally, the rest of the world, the poorest 25% of the emitters, starts reducing their emissions beginning in 2040 and following the same pattern. The logic behind the two phases is that there are simpler or wasteful uses of energy that can and should be eliminated faster, while other needs might take longer to eliminate or find substitutes for. Under this scenario, the CO_2 concentration in the atmosphere reaches 505 ppm by 2100.

The third scenario, labeled "Linear 80y", assumes a linear reduction of the emissions during an 80-year period, with the same proportions and timing for the groups, with developed countries starting in 2020; less developed countries starting in 2030 and the poorest countries starting in 2040. Under this scenario, the concentration of CO_2 in the atmosphere reaches 478 ppm by 2100.

The final scenario "Linear 50y", is similar to the previous one, with the difference that the linear reduction of emissions is now over a 50-year period. Under this scenario, the concentration of CO_2 in the atmosphere reaches 456 ppm by 2100.

The world would continue emitting CO_2 in the year 2100 in all but the last scenario. Under the BAU scenario, the world would be emitting 77.5 GTY in 2100, under the Combined scenario 13.6 GTY in 2100, and in the Linear 80y case, 7.26 GTY in 2100. It would stop emitting completely in the Linear 50y scenario.

The logic behind the breakdown between developed countries, less developed countries and poor countries and the staged implementation is simple. The developed countries are the worst emitters and can more easily afford the initial high costs of re-newable alternatives, hopefully resulting in a lower cost in the future. Less developed and poor countries have lower standards of living and might need time to catch up with what we in the developed world consider basic necessities. Just like some less developed countries bypassed building a network for landline telephones, going directly to cell phones, hopefully the less developed countries can bypass fossil fuel generation of electricity and move directly into renewable alternatives.

Other scenarios can be envisioned. The goal is the quicker and less expensive, but also the fairest way of arresting the emissions of CO$_2$. By arresting CO$_2$ emission, I mean totally. We have overcome the capabilities of Mother Nature to recycle the CO$_2$. It is not a matter of reducing it 20% by 2020, or 30% by 2030, or 75% by 2075. It means reducing emissions 100% as soon as possible. Even 50 years is a long, long time, and if feedback mechanisms exist, there might be a point of no return. We have to make the effort to not burn any more fossil fuels at all, nothing, zilch, rien, nada.

I am aware that with the previous paragraph I might have lost the respect of most readers. How naïve! He was probably high on something when he wrote that! How can we possible give up all the advances and comforts we have achieved permanently, forever? Doesn't he know that without fertilizer there will not be enough food for everyone? That without electricity there will be no computers, knowledge, problem solving? That without gasoline to distribute products, food will rot in some places and there will be hunger in others?

I am aware that we simply cannot stop cold turkey. The economy and civilization would collapse. There has to be a transitional period. The shorter, the better. By signaling our resolve in the US, we can send a clear message to the oil producing countries, to the public utilities, to friendly and not-so-friendly countries and even to the public. We also need to work on the international front to get everybody on the same page because we need the whole world to participate; otherwise the sacrifices made by those who were brave enough to recognize the threat will only postpone the inevitable

The transition period is not while we restock the horse population (I have nothing against horses!), but for the period when the cars with internal combustion engines would be replaced by fuel cell or electric cars and the hydrogen infrastructure is developed, new super efficient houses and buildings are built which would be requiring much less energy to maintain at a comfortable level and we find ways of producing bio-plastics. It is not intended to be a gradual decline into medieval times, but a transition period into a carbonless society.

Quite a challenge, isn't it? Let's hope that those politicians, diplomats, scientists and engineers work fast.

Chapter 2 The Problem

The problem is huge and multi-faceted. Many of the obstacles will be discussed in detail in subsequent chapters. There are technical problems. There are economic problems related to the high investment required which would translate into higher costs for almost all items (inflation). There are also issues related to the financing of the task. Finally, there are human problems, related to how to proceed, convincing the opposition and affected persons here and in the rest of the world.

Eliminating CO_2 emissions appears to be an insurmountable problem. We cannot burn fossil fuels without producing CO_2 and we cannot sequester it (despite claims to the contrary by effusive proponents of the best example of an oxymoron - Clean Coal!). Everybody would quickly endorse a solution that is better than burning fossil fuels, provided it is cheaper, but very few are prepared to pay more. Unfortunately, there is no magic solution. We know that higher energy prices will create havoc in the economy, that the technical solutions available today are not perfect and are unfortunately quite costly, but our leaders have failed before and are failing us now, because rather than leading us, they are waiting for a miracle.

All scientists are in agreement that CO_2 traps heat, and therefore an increase in CO_2 concentration would result in a warmer planet. A few of them disagree with the conclusion that we have overwhelmed the natural mechanism for recycling CO_2, proposing instead that plants might thrive on a rich CO_2 atmosphere, which might level the CO_2 growth. There is also vociferous opposition from other groups, the "deniers" dismissing the findings and methodologies used by the scientists or pointing out that they are failing to take into account the influence of the Sun's cycles or known periodic perturbations of the Earth's orbit. The deniers add another level of confusion to a complicated problem, and our elected leaders keep waiting for a consensus to be reached, purposely forgetting that there are very few legislative acts that had been approved by unanimity.

Worse than overt opposition is the passive attitude. The human brain is not designed to trigger a response to a long-term future threat; most people are optimistic and believe that technology is going to provide a solution; the alternative of not burning fossil fuels would likely cause a collapse of the economy, and there is serious possibility that our standard of living would revert to medieval times; no politician suggesting placing a limit on burning fossil fuels would be re-elected, and international discussions are bogged down with blame and recrimination aimed at developed countries to exonerate less developed. The Kyoto Protocol, reached in 1997 set a modest target of reducing CO_2 emissions to 5% below the emissions of 1990. Even among the signatories, the target has proven elusive.

A good justification for the inaction is the complexity of the problem, which includes technical difficulties with renewable energy sources, the size of the endeavor, political realities, international law and enforcement, fairness, cost, financing mechanisms, the time required and opposition from numerous parties, depending on the issue.

Technical Problem

The technical problem with renewable energy has to do mostly with its intermittency and cost. Photovoltaic solar panels only work during the day, and during the winter it will not be possible to watch the 5 o'clock news using PV panels. Wind is not as steady, reliable or predictable as we would like. We must also find a better way of storing energy. Batteries are bulky, heavy, expensive, and last only 5 - 7 years. The high cost is the result of two irrefutable realities: (i) solar radiation is huge, but it is extremely diluted, and therefore requires large surface areas to capture a meaningful amount of energy, and; (ii) it is constantly changing; there are long days of sunlight and short days, there are times when the Sun rises high and times when it stays low. The plants know that. Most of them shed their leaves and go dormant in winter.

Size of the Endeavor

The size of the endeavor is not obvious at first, at least for us in the US, accustomed to the market providing incentives so that we can start enjoying renewable energy without too much disruption or inconvenience and only a modest cost increase. It has not taken that long to have Internet services or cell phone coverage almost everywhere. But,

1. The world energy consumption in 2014 was almost 13 billion tons of oil equivalent ("toe"), which is about 1.8 tons per person per year or about 5 kg of oil equivalent per person per day.

2. The world consumes about 92+ million barrels of oil per day (14.6 million cubic meters per day or about 13.2 million tons per day or the flow rate of the Mississippi river at New Orleans in about 15 minutes), which almost magically gets transformed into refined products that are available for sale almost everywhere. To transport the oil, very large oil carriers, some 400 m long, displacing up to 500,000 tons and carrying millions of barrels, circle the world day and night and are responsible for about a third of ocean freight. To put it in perspective, only regarding primary energy, with the world's population today at 7.2 billion people, there is a need to transport about 1,000 lbs of oil and oil products plus about 2,300 lbs of coal (average energy density 2 tons of coal/toe) and deliver about 16,700 cubic feet of natural gas for every living person on this planet per year. The investment in the world's infrastructure, ports, pipelines, storage tanks and even dispensing stations is trillions of dollars.

3. The world consumes about 22 PWh/y (22 $\times 10^{15}$ Wh/y or 22,000 billion kWh/y) of electricity or about 3,000

kWh per person per year. The distribution infrastructure, a patchwork of new and old technologies, is likely to be the basis for the renewable energy infrastructure and should remain in place, but most of the generation equipment (about 4 million MW installed capacity) might need to be replaced. The replacement cost of the generation equipment alone is several dozens of trillions of dollars.

4. There are almost one billion cars in the world with an average lifespan of about 15 years, which are going to need fossil fuels for many years. Replacing all those cars with electric or fuel cell cars is going to take a couple of decades, but that means the demand for electricity is going to increase two or three times in the future.

Political Problems

The political problems are plentiful. Leaders need to project an aura of confidence and therefore prefer to talk about a bright future clinging to perceived discrepancies in scientific studies, promising alternatives like biofuels from switch grass, or that the hydrogen economy is just around the corner and the little details will be worked out or possible advances in nuclear fusion. The oil and coal lobbies are extremely powerful and have dismissed the threat of global warming as a hoax and used their massive resources to seed "valid" concerns. A typical answer to questions about climate change is: "I am not a scientist". Jimmy Carter's speech in 1978, appearing in a cardigan and suggesting lowering thermostats, was ridiculed and probably cost him his re-election. Politicians know that if they talk about reducing energy consumption, it will be translated into reducing the standard of living, recession, unemployment, etc. They have the lobbyists and constituents to worry about. So, they take the safest course available, which is no action.

International Problem

International law and enforcement is another difficult problem to tackle. Obviously developed countries started earlier emitting CO_2, emit the largest amount per capita, and bluntly, they are responsible for a large portion of the accumulated CO_2 in the atmosphere. Less developed countries started later, and even though their per capita emission is less, China, with the largest population, is now the largest emitter. The problem is that CO_2 emitted by Argentina is as bad as CO_2 emitted by Zambia, and unless the world acknowledges, cooperates and agrees to reduce emissions, the problem will not get solved. That opens a can of worms. Should less developed countries be grandfathered in until they reach the same level of emission as developed countries? Should developed countries finance renewable energies for less developed countries as reparation for the harm inflicted? How would it be enforced? Is the US going to be dictated to by the UN and told that its contribution this year is x billion dollars? Are developed countries legally responsible? Lawyers would spend eons debating whether lack of knowledge of the consequences is sufficient grounds to dismiss the charges. Do we have an obligation to future generations? Or put it in other ways, do we have to make sacrifices so that they inherit a livable planet?

Fairness Problem

The issue of fairness needs to take a central role in any discussion. You cannot tell the 1-1.5 billion people without access to electricity "we are so sorry you do not have electricity and cannot read at night, or keep your food fresh, watch TV or be part of the digital world, but generating electricity emits CO_2". The issue of fairness also has to do with the allocation of pain, with those emitting more than others having to make bigger sacrifices than those who do not have a car or air conditioning in their houses. To the chagrin of many, we cannot let market forces run free, because the golden rule has a corollary that is unfortunately true, that "those that have the gold make the rules".

Cost and Time

Cost and time are intertwined. It can be done at an affordable pace, as was done with the introduction of electricity, the telephone and water distribution systems, replacing older and more polluting plants first. If it takes 100 years, so be it. It can be done at a faster pace, but then we are replacing equipment and need to allocate massive resources. The cost is staggering, and by that, I really mean incredible. It is not only a matter of replacing generating equipment MW per MW, but installing more equipment because some of the intermittently generated energy needs to be stored for cloudy or windless periods plus we need to use renewable energy for transportation. If we were to substitute all primary energy consumption in the US with wind turbines (23% availability) the number of 1 MW wind turbines required would be about ten million. With 50% storage efficiency, the number doubles again. The astronomical cost now has to be tripled to add the cost of the storage devices. We are talking of tens of trillions of dollars for the US alone. The total cost for the world could reach a couple of hundred trillion dollars. That is a very good reason for inaction and foot dragging.

How much time do we have? It is just a matter of deciding what level of CO_2 to aim for. If we continue with business as usual, the level of CO_2 will reach 440 ppm in 2030; 490 in 2050 and 550 ppm in 2070. If we just do not increase emissions, it will reach 435 ppm in 2030. If we try for 50% by 2030 (assuming linear reduction), the level will reach 426 ppm by 2030. If we aim to reduce emissions by 50% in 2050, the level will reach 465 ppm in 2050; if we target to reduce emissions by 70% in 2050, the level will reach 457 ppm. Conclusion: we have little time.

Implementation Problems

As will be discussed later, there are only five possible mechanisms to arrest the emission of CO_2. The first two are wishful thinking: voluntary and regulation. The former needs constant reinforcement and fades quickly, especially if it gets hot, or cold.

After some hours of misery, you would go to your thermostat and put it back to where you like it. Regulations are fine, but you cannot force electric utilities to invest when they do not have the balance sheet to satisfy lenders or regulate that everybody have electric cars by 2020. You can only regulate something that is doable, within the capabilities of the people.

That leaves three other alternatives. The preferred choice of many is cap and trade, letting the market decide the methodology and price of emission rights, with some parties or countries planting trees to sequester CO_2 and others making massive installations of solar panels. The carbon tax is probably easier to implement. The customer ends up paying additional monies for goods, be it gasoline, electricity or strawberries from California that were transported by truck. Gas stations and electric companies are already tax collectors. Carbon rationing is probably the best mechanism because it is fair and would be politically more palatable to implement. Every human being would be entitled to and is given transferable rights to emit the same amount of CO_2. This amount of CO_2 would be reduced over time.

The carbon rationing mechanism could be applied worldwide, with countries emitting less CO_2 having the right to sell and countries needing rights acquiring them. In developed countries with sophisticated banking systems, the free market could be used to define the current price of the rights to emit, with those needing them buying rights from those who are more frugal. On the international level, an agency could coordinate the transfer from countries with surpluses to countries with deficits at a fixed rate that would allow less developed countries to finance their transition to a carbonless society.

The road is long and arduous, full of discontent and pain, with prices of many items skyrocketing beyond the reach of many, causing inflation and destabilizing today's economy into a new one that would have to be invented as we go along. But it is also full of opportunities, because new industries will be created offering plenty of local jobs.

Those opposed to the changes, those who believe that it is a hoax or a communist scheme to take the wealth from the rich, generally known as "deniers," will use their resources and intelligence to block progress, accusing those proposing change, generally known as the "catastrophics," of spreading false fears and lies.

So far, all prophets of doom have been wrong, and I do not see any major benefits to adding my name to that list. My efforts are to shed some light into possibilities. I hope that a discussion on the carbon rationing mechanism will lead to constructive discussion among less developed and developed countries, among political parties, whether they are Republican or Democrat, Labor or Tory, socialist or right wing. I also hope that my plea for a massive research effort is well received and that industry, universities and governments provide the resources to explore many possible avenues and/or solve many of the nagging problems and limitations of the current alternatives.

Finally, I hope that we react quickly, and that my children, grandchildren and their descendants will have the opportunity to live in a prosperous world, with newer and better gadgets, health, wealth and peace, and that one day, historians will recognize that our generation had the courage to take the needed steps to stop emitting more CO_2.

Chapter 3 Energy

The availability of fossil fuels has allowed a tremendous increase in the world population, but also has contributed to a significant improvement in the standard of living. We live in wondrous times, no longer limited by what is available locally. Instead, the global economy allows us to have orange juice year around, copper from Peru, tungsten from Bolivia, flowers in December from Colombia and shirts made in China with cotton produced in Egypt and plenty of food. We can remain comfortable during cold winters and hot summers. Granted, not everyone has benefited at the same pace, but the large majority has experienced marked improvement in their lives.

Nice words. Maybe a poem to energy would be appropriate. But burning fossil fuels produces CO_2. This chapter deals with a summary of the size of the fossil fuel market and the effect it has on the GDP as a measure of wealth.

Life is an endo-energetic process. The further up you are on the ladder, the more energy you need. Taming fire allowed mankind to evolve. The high energy density of fossil fuels transformed us. We have become addicted to energy in a big way.

The link between energy consumption and world population[4] is shown in Figure No. 9. The world population has grown from about 200 million, 2000 years ago, to 500 million around the time of the discovery of America to about 7.2 billion today. Energy consumption has gone from muscles and burning wood to about 13 billion tons of oil equivalent. Translating such big numbers that are meaningless to most people into more common units, we can say that today, on average, every person consumes 1.8 tons of oil equivalent per year or about 1.4 gallons of gasoline per day (a bit less due to efficiency issues) or about 1.9 kWh every hour of the day. That energy is used to produce steel, cement, glass, aluminum, to construct housing, to provide light, heating and

---

[4] Ehrlich PR, Kareiva PM, Daily GC (2012) Nature 486(7401):68–73.

cooling, for communications, to move products from faraway places to our local supermarkets and stores, for fertilizers, insecticides, plastics and medicines that keep us healthy and living twice as long as people did during the Renaissance.

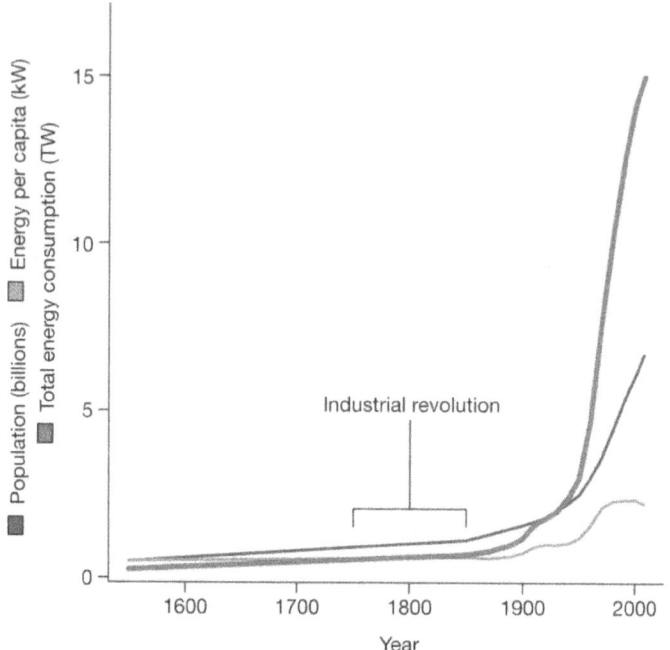

Figure 9 Energy and Population

The energy is derived from four main sources (primary sources): fossil fuels, nuclear energy, hydroelectricity and renewable energy, with fossil fuels further subdivided into oil, natural gas and coal. The total primary energy used by the world in 2014 was 13 billion toe (it does not include some non tradable renewable sources like wood or dung gathered by the poorest members of humanity as their source of energy). Renewable energy refers only to tradable, modern use, like solar, wind or biofuels.

Last year, fossil fuels represented 86% of the primary sources, and all emit CO_2 when used. Even in a hypothetical best case scenario for a prompt and orderly transition to a carbonless society, its use would be necessary during the transition period. Fortunately, there are sufficient reserves, and the transition can occur because we choose it, not because we run out of fossil fuels.

Total oil world reserves, as provided by the 2014 BP Statistical Energy Review, are estimated to be 1.701 trillion (10^{12}) barrels, with a reserve to production ratio of 50.6 years. Of this amount, 1.216 trillion barrels are in OPEC countries, representing 71.6% of the world's oil reserves (a detailed discussion of oil reserves will be presented in the next chapter).

Total natural gas world reserves are an estimated 6,606 trillion cubic feet, with a reserve to production ratio of 54.3 years. Almost 50% of the reserves are located in three countries: 25.2% in Russia, 18.2% in Iran and 13.1% in Qatar.

Total world reserves of coal are estimated to be 892 billion tons, with a reserve to production ratio of 110 years. Of this amount, 237 billion tons are in the US, representing 26.6% of the total, with a reserve to production ratio of 262 years.

For many, the problem is solved. There is plenty of energy to carry us until the time a non-polluting, non-CO_2 emitting source of energy can be developed. There is oil and gas for half a century. Let scientists and engineers come up with the solution.

Energy Consumption per Capita

A comparison of the energy consumption per capita in 2011, expressed as tons of oil equivalent per person-year for a few selected countries is shown in Figure No. 10.

Several factors must be taken into consideration to properly understand the graph:

- energy producers must spend energy extracting resources, but also tend to have very low local prices for energy, which translates into disproportionate use;
- countries with colder weather tend to have a relatively higher use of energy because they need more heating;

- the huge disparity in energy use primarily reflects different income levels throughout the world, with higher income countries able to afford more energy.

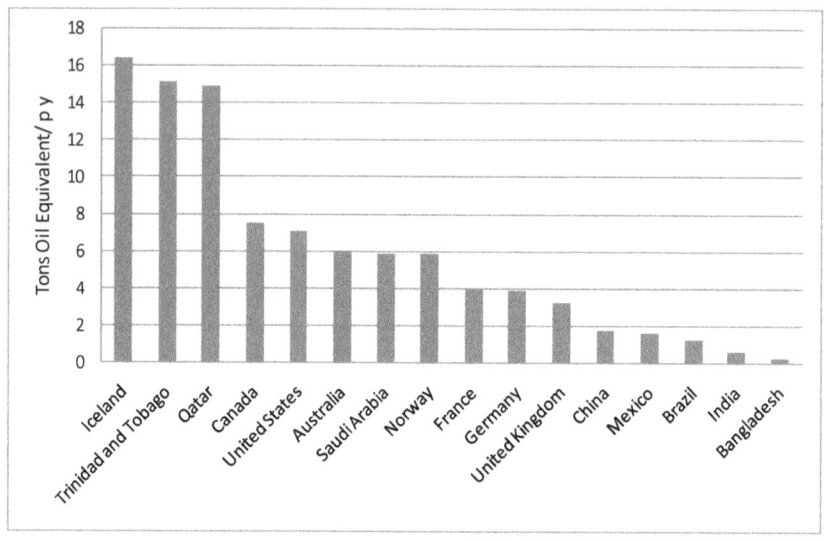

Figure 10 Energy per Capita Consumption

The country with the highest energy use per capita is Iceland, with a large portion of inexpensive geothermal energy from locally occurring volcanic hot springs. In 2011, by coincidence, the per capita consumption of the U.S. was 4.2 larger than the world average and also 4.2 times larger than the consumption of China. Comparing only tradable energy (i.e. assuming that nuclear energy, hydro power and coal are local resources and thus not tradable), the consumption per capita of oil and natural gas in the U.S. was 4.9 times the average world consumption and 8.9 times larger than the per capita consumption in China.

However, when we compare the consumption of energy as a percentage of the GDP, the picture is quite different, as shown in Figure No. 11.

The graph was constructed utilizing data from the BP Statistical Energy Review for the breakdown of energy (expressed in million tons of oil equivalent), the IMF Financial Statistics for the GDP and a value of $70/bbl for oil. Using another price of oil would change the value of the ordinate, but the order and relative size of consumptions would remain constant. With energy consumption as a percentage of GDP, the U.S. is in the middle of the pack. There are a couple of reasons:

- countries with few local energy resources need to spend more on oil imports and thus have developed more efficient, frugal systems;
- more developed countries sell more services (consuming less energy than in producing industrial goods);
- abundance of local resources (whether it is coal in China or oil and natural gas in Russia) allows governments to either subsidize energy prices or disregard the need to improve efficiencies, resulting in more consumption, and;
- countries with large GDPs spend a smaller percentage of their resources on energy, similar to rich families that spend a smaller portion of their budget on gasoline.

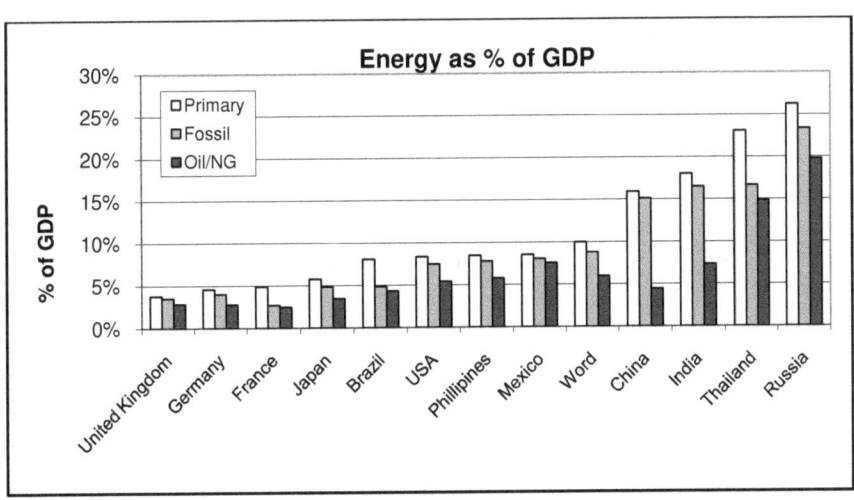

Figure 11 Energy as a Percentage of GDP

Figure No. 11 shows that the world, on average, spends 10% of its GDP (at $70/bbl) on energy, while the U.S. spends only 8%

and some European countries less than 5%. The implications are that more advanced countries are likely to feel less pain than the rest of the world to an increase in the price of energy.

Energy and Free Trade

Energy is an important component of almost everything:

- To produce food, we rely on fertilizers. One ton of ammonia, the building block of all nitrogen fertilizers, requires about 35 million BTU, which is equivalent to about seven tons of natural gas. The price of ammonia increased almost fivefold, along with the price of oil, then dropped drastically during the financial crisis and is now hovering around two times the previous decade's average. As a result, the price of most commodities (corn, wheat, etc.) went up to cover the additional cost of the fertilizer or to quench demand, given lower production due to a lack of fertilizer.
- T-shirts, tennis shoes, etc., imported from China and other countries require fertilizer for the cotton, plastics for the soles, energy to produce the fabrics, to sew the pieces, for packing and transportation.
- China imports cotton from Egypt, wool from New Zealand, petrochemicals, minerals, specialized electronics and many other materials and components to manufacture finished products.
- The allocation of minerals is not evenly divided amount countries, and we have to rely on importing minerals from mines located all over the world. Minerals require energy to extract from mines, concentrate them and convert them into billets or ingots.
- Fossil fuels are necessary for transporting goods all over the world and air travel requires high density energy fuels.
- The Internet requires electricity to power servers holding the information and the computers retrieving it.

The price of many commodities, from metals to agricultural products, touched historical highs when the price of oil peaked, reflecting the higher cost of fuel. For a while, companies can absorb the price increase, but sooner or later the increased cost gets reflected in the price of the product. The public, with less disposable income due to the high price of gasoline, starts curtailing purchases, first of non-essential goods and then eventually seeks even cheaper substitutes for food and staples. Demand for products decreases. Higher prices without increasing demand produces stagnation and then price decrease with deflation.

If exporting countries experience weakened demand for their exports, they have less foreign currency to purchase raw materials, food and energy. Unemployment soars. Free trade is likely to shrink with high oil prices. The U.S. imports flowers from Colombia that are flown in daily, but as the cost of the flowers goes up due to the rising cost of fertilizer and jet fuel, the buyers with less disposable income buy fewer flowers. For Colombia, importing fertilizer and jet fuel, this translates into fewer exports.

Price Elasticity
Economic theory states that demand for goods and services gets adjusted by price. If the price goes up, demand shrinks and conversely, if prices decrease, demand increases. Price elasticity is a measure of the sensitivity of demand change to price changes. When the change in demand is smaller than the change in price (elasticity less than one), the price sensitivity is considered inelastic. When the price elasticity of demand for a good is elastic (elasticity greater than one), the percentage change in quantity demanded is greater than that in price. With inelastic demand, producers earn higher revenues with reduced demand, giving them little incentive to lower prices.

Several factors affect the price elasticity of a product:

- Substitutes: The more substitutes there are for the product, the higher the elasticity, as people can easily switch from one good to another if a minor price change is made. For gasoline, there are no substitutes. Some modifications can convert a car to use ethanol or one can buy a new diesel car. There are also no substitutes for electricity.

- Necessity: The more necessary a good is, the lower the elasticity, as people will buy it no matter the price. Gasoline is necessary. For many people living in the suburbs, there are not many alternatives (no public transportation, grocery stores are miles away, etc.). We can survive a couple of hours without electricity, but life becomes miserable after a week without it.
- Percentage of income: The higher the percentage that the product's price is of the consumer's income, the higher the elasticity, as people will be more cautious about purchasing the good because of its cost.

Demand for energy products has been inelastic in the past. The price of gasoline has doubled in the last decade, yet consumption has only decreased a few percentage points.

Elasticity allows producers to attempt to optimize their revenue stream. The longer the price stays within a small range, the more sensitive people are to price increases, but once they get over the shock and are forced to accept it because it is needed and there are no substitutes, producers can maintain the new price with little concern for shrinking demand. A good example is the recent surge in gasoline prices. When oil hit $140-150/barrel, gasoline prices surged to $4/gallon. When oil prices decreased to $70/barrel, one would have expected gasoline prices to return to the $2.00/gallon range. The producers, sensing that a sensitive price barrier had been broken, only reduced the price to about $2.70 – 3.00/gallon.

Economic theory assumes that high price will foster production. So far, high prices of oil in the '70s resulted in additional production; the high prices in this century allowed production of pricey unconventional oil in Canada and the US. But, as expected, it has dampened demand considerably. With prices of most things going up because of high oil prices, many shifted their expenses to subsistence items and eliminated or drastically curtailed their driving, resulting in a substantial drop in demand, which has forced the price of oil down to about $50/barrel. Since the new unconventional oil is more expensive to produce, at $50/barrel it is uneconomical. Drillers might encounter difficulties securing financing for new wells, eventually reducing output.

Energy in the US

During almost all of the last century, the US was the largest consumer of energy in the world, in absolute and even in per capita terms. The US has been fortunate to have plenty of oil, natural gas and coal, and also fertile soil and plenty of water, flanked by two oceans and secure borders, with political, legal and educational systems that fostered a free market economy, independence and entrepreneurship.

Table No. 2 below shows for 2014 the U.S. and world consumption of energy, broken down by primary source and the corresponding level of reserves of each primary source. Other, minor sources not traded (mainly wood for heating or cooking in poor areas) are not included. While the U.S. consumes 19.9% of the oil, its reserves (48.5 billion barrels) represent only 2.9% of the world reserves and would only suffice to cover demand for 11.4 years at the present extraction rate. The US, with 4.4% of the population, consumes 17.8% of all primary energy. The rapid increase of energy consumption in China has lowered the share of total energy consumed by the US, which at the beginning of the century represented 25% of the total.

While US consumption of primary energy has been rather stable (it peaked at 2,371.7 million tons of oil equivalent "mtoe" in 2007, dropped to 2,205.9 mtoe in 2009, with an average of 2,291.3 mtoe), world demand has steadily increased from 10,916

41

US Primary Energy Consumption and Reserves							
(million tons oil equivalent)							
Source	U.S.A	World	(%)	Reserves	(%)	R/P	Units
Oil	836.1	4,211.1	19.9%	48.5	2.9	11.4	billion barrels
Natural Gas	695.3	3,065.5	22.7%	345.0	5.2	13.4	trillion cuft
Coal	453.4	3,881.8	11.7%	237.3	26.6	262.0	billion tons
Nuclear	189.8	574.0	33.1%				
Hydroelectric	59.1	879.0	6.7%				
Renewable	65.0	319.6	20.3%				
Total	2,298.7	12,931.0	17.8%				

Table 2 US/World Primary Energy Consumption/Reserves

mtoe in 2005 to 12,918.7 mtoe in 2014, with China showing the largest gain, going from 1,791.4 in 2005 to 2,972.1 mtoe in 2014, and India going from 366.8 mtoe to 637.8 mtoe in the same period. Electricity is not a primary source (except for hydroelectricity). It is produced mainly by burning coal or natural gas or by nuclear energy.

Despite the drastic increase of production of shale oil, the U.S. still imports 8-9 million barrels of oil equivalent per day. At $70/barrel, this still represents $0.6 billion per day or $210 billion per year, or about $700 per year per person ($2/day per person).

Energy is interchangeable and can have multiple uses. Natural gas is used to generate electricity, but also for heating and cooking. Heat pumps using electricity provide heating and cooling.

Starting with the primary sources of energy, the U.S. consumed in 2014 2,298 million tons of oil equivalent or 7,200 kg of oil equivalent per year per person or 20 kg of oil equivalent per person per day or about 4.75 gallons of oil per person per day, which at the recent price of $60/barrel, represented a cost of primary energy per person of $7/day.

Attempting to simplify matters to typical consumption of electricity and gasoline would require assuming conversion efficiencies and yields of gasoline per barrel of oil and allocation of other items that we might not purchase directly (most of us do not buy diesel fuel directly, but pay for the transportation of products; we do not buy plastic, but products made out of plastic; etc.). The approach followed might appear simplistic to knowledgeable readers, but avoids questions of double counting or validity of conversion efficiencies. About 75% of the oil is transformed into light and medium distillates used for transportation, and most of the coal, nuclear energy and hydro power is used for electricity. Renewable energy is used mainly to generate electricity, but in the US, biofuels are important (corn for ethanol), so we are going to allocate 50% to transportation and 50% to electricity. The only assumption that we have to make is a breakdown of the use of natural gas for heating, industrial use and generation of electricity. During the winter, about a third of the natural gas consumption is for residential use while in the summer it is only about 10%. We will assume that 20% is used for heating, 40% for electricity generation and 40% for other industrial uses.

Translating the above table of primary energy sources into a table of uses of energy per person (with the caveat that electricity is also used for heating and cooling), is shown in Table No. 3:

US Per Capita Energy Consumption					
(kg of oil equivalent per day)					
Source	Mill T/y	kg/p d	kg CO_2/p d	Use	kg/p d
Oil	836.1	7.16	24.74	Transportation	5.65
Natural Gas	695.3	5.95	18.01	Other uses of Oil	1.79
Coal	453.4	3.88	15.66	Heating	1.19
Nuclear	189.8	1.63		Other uses of NG	2.38
Hydroelectric	59.1	0.51		Electricity	8.67
Renewables	65.0	0.56			
Total	2,298.7	19.68	58.41		19.68

Table 3 US Per Capita Energy Consumption

With the translation, we have simplified the primary sources into transportation, heating, electricity and other uses. We have separated other uses of oil, like petrochemicals or asphalt, expressed as kg of oil equivalent (koe), which represent about 1.79 koe/p d and other uses of natural gas, used in manufacturing fertilizers and in a myriad of other industrial processes, which are about 2.38 koe/pd, representing 4.17 koe/pd that could not be rapidly replaced from renewable energy sources. The remaining 15.51 koe/pd is the amount that could be substituted for by non-CO_2 emitting renewable energy sources if we are to arrest the buildup of CO_2 in the atmosphere. The breakdown allows us to define priorities and the effects on our daily lives.

In summary, the consumption of fossil fuels has resulted in a population explosion and a surge in global trade worldwide. Energy has shown very little elasticity, allowing a massive transfer of funds from the rest of the world into the hands of oil producing countries.

The US was for many years the largest emitter of CO_2, both as a country and on a per capita basis. Today, the US emits approximately seven Gtons/year or about 20 tons per person year or approximately 58 kg/p d.

.

Chapter 4 The Fossil Fuel Era

To put the CO_2 emission problem in proper perspective, this chapter will superficially discuss the availability of fossil fuels. There is a plethora of books about oil and other fossil fuels, some giving a historical perspective, geology and the effects of booms and busts on towns, while others discuss their impact on civilization, geopolitics and even wars. Below is a quick synopsis, concentrating on oil, the most versatile fossil fuel.

The rapid advance of the Industrial Revolution at the beginning of the1800s was made possible by the utilization of coal as a source of energy. Coal's main advantages over wood were: (i) it has a higher energy density; (ii) it was concentrated in mines rather than distributed around the country side, and; (iii) it reached higher temperatures. Consumption initially for heating and cooking was quickly surpassed by industrial use to generate steam to move a variety of pumps and other machinery. It did not take long for entrepreneurs to discover that coal could also be used in transportation, with the development of the steam locomotives and steam ships.

Around 1859, oil was first pumped out of the ground in Pennsylvania and quickly supplanted coal as a more versatile fuel because, as a liquid, it could feed engines by gravity, eliminating the need to shovel coal into furnaces. Consumption of oil grew gradually and steadily. The US was fortunate to have plenty of oil deposits, and up to the 1950s, was the largest exporter in the world. The importance of oil became critical during the Second World War. Fighting between the Allies and Germany took place in North Africa to control the Suez Canal, the flow of oil and raw materials from the Far East. Following the war, the US developed strategic alliances with Arab countries to guarantee access to oil. During the 50s and 60s, the world experienced a period of abundant and inexpensive oil.

Oil is a very versatile source of energy. Through distillation, it is separated into a range of components that have different physical properties and fill various niches. The lightest components – methane, ethane and propane – are first separated. Methane is the main component of natural gas, ethane is the building block for the petrochemical industry and propane is the main component of liquefied natural gas (LNG). A second group of products forms different fuels (gasoline, diesel, jet fuel, etc.), while a third group contains a mixture of valuable products utilized by industry.

Most of the oil is used as fuel. Gasoline is used for transportation in internal combustion engines (ICE), given its high storage density and quick flammability. Diesel is used mainly in heavier transportation modules (trucks, locomotives, ships). Heavier fuels are used for heating and generation of electricity in large thermoelectric plants.

The oil embargo of 1973 was a wakeup call for the world, and particularly for the US, about the dangers of relying on imported oil to maintain its oil consumption habits. The government's response was quick and effective and, among other matters, eventually introduced a fleet gas efficiency requirement, publishing the Corporate Average Fuel Economy (CAFE) standards. The embargo also brought about a deluge of research into alternative sources of energy. The Department of Energy directed efforts into all conceivable forms of easing the need for imported oil. There was a surge of patents granted, hundreds of wind turbines were installed in Altamont Pass in California and the construction of the first solar energy generating systems (SEGS) began by the end of the decade.

The high price of oil had a positive repercussion. It justified further exploration and exploitation of more difficult areas. As a result of the additional investment, oil started flowing abundantly, and by the middle of the 1980s, the price of oil had abated to less than half its peak in inflation-adjusted terms. The declining price removed the urgency for renewable energy. Oil was again cheap and abundant, so there was no need to continue struggling to improve efficiencies to be competitive. Research funding and tax incentives for alternative sources of energy dried up.

Fast forward 20 years, where we found ourselves basically in the same situation. The price of oil reached $140/barrel in July 2008, breaking the 1974 price peak on an inflation-adjusted basis, with the US importing the equivalent of 14-15 million barrels per day, at a cost of $1.5 billion per day. However, the high price made shale oil feasible, and today, for the first time in 30-40 years, the US now produces more oil than it imports, and some are stating that the US again will surpass Saudi Arabia to become the world's largest producer.

Oil Reserves

Analogous to cherry picking, all major oil fields have been discovered. Early production was from rather superficial wells (low-hanging fruit), and as prices and technology developed, additional easy-to-exploit cherries have been picked. With the understanding of the geological requirements for the existence of oil deposits, the field of exploration narrowed. Although not all areas are accessible for assessment due to political considerations (some countries do not allow foreign exploration), it is rather safe to predict that there are no major deposits left on land. With the North Sea, the Campos basin in Brazil and the Nigeria/Angola deposits, major finds to depths of about 3,000 meters have been made. There might be some oil deeper down, but they are not going to be massive deposits.

Although we often see reports of new major oil finds, the term "major" is relative. Saudi's Ghawar field, discovered in 1948, supposedly held 87 billion barrels. In the year 2000, there were sixteen oil discoveries of giant oil and gas fields of 500 million barrels of oil equivalent or bigger. In 2001, there were nine. In 2002, there were two, and in 2003, none[5]. Again, high oil prices justified more exploration, and in the last few years, there have been other finds of conventional oil.

---

[5] The Empty Tank – Jeremy Legget – Random House 2005.

At the beginning of the 21st century, reserves stood at about 1.3 billion barrels. By 2014, reserves had risen to 1.701 trillion barrels (1.701 x 10^{12} barrels), with about 200 billion barrels added by Venezuela, another 131 billion barrels by other members of OPEC and only some 45 billion barrels by the rest of the world[6]. Most of the available reserves are located in the Middle East, and OPEC has about 70% of the worldwide reserves. However, despite the importance, those published numbers need to be taken with a grain of salt:

- The size of the oil reserves is a complicated business. Reserves are broken down into three categories: proven, probable and potential, with the caveat that they are recoverable based on cost assumptions. The proven reserves are based on statistical analysis of a few test drillings made to assess the size of the field, and assume a homogeneous depth or thickness of the field.

- Furthermore, not all the information is verifiable. Some countries will not allow international audits to analyze their data, and while in the US there are standards used to assess reserves, not all countries adhere to those standards. For years, BP published an oil reserves report (BP Statistical Review of World Energy), but after the embarrassment of Shell having to lower their reserves figures, BP now includes a footnote stating that: *"the estimates have been compiled from a variety of primary official sources and third party data from the OPEC's Secretariat... the data does not necessarily meet the SEC definition and guidelines for determining proven reserves... and it does not necessary represent BP's point of view of proven reserves by country"*. In the mid 80s, certain member countries of OPEC unilaterally increased the size of their reserves (just to increase their allotted quota), claiming the previously stated figures were conservative.

---

[6] Oil and Gas Journal – 2014 Worldwide Production Report

- If it is true that the reserves are based on their being recoverable at a given price, the amount of reserves should change often with oil prices, but despite swings from $20 to $140 and then back again to $60/barrel, OPEC's total stated reserves remained constant.
- Finally, reserves should be reduced periodically by subtracting production, but apparently OPEC does not feel the need to conform to SEC's accounting standards.

Another useful measure is the reserves to production ratio, usually expressed in years of operation at the current rate of extraction. It adds another complication. Stating that there are 40 years of reserves at the current rate of extraction means that there are only 29 years of reserves at an extraction rate that grows yearly by 2% (the average growth rate of oil consumption).

The evolution of oil reserves is carefully followed by many experts worldwide, and new finds or depletions are quickly reflected in the share price of oil companies. There is no consensus on available reserves. On one hand, the "optimistic" group, composed of most oil companies, governments and their agencies, financial analysts and business journalists tells us that there are about two trillion barrels of oil to be found and recovered, therefore there is no need to be concerned. There will be plenty of oil for your children, their children and their grandchildren. That level of reserves was true before fracking (fracturing - method to extract oil from shale deposits), which has just changed the game, and now there are several trillion barrels all over the world, with one trillion in the US. The "pessimistic" group, on the other hand, believes that there is less than one trillion barrels of oil left, and while we have been successful in extracting oil from shale deposits, it needs to be priced about $100/barrel to be profitable.

Worldwide demand for energy will continue growing in the near future as a consequence of population growth and increases both in the level of development and income per capita of the population. Historically, worldwide demand has been growing at about 3.5% per annum. At this rate, demand for oil will double in approximately 20 years from the current consumption rate of about 86 million bpd to 172 million bpd.

If most of the easy cherries have been picked, it appears unlikely that we will have a second or third "harvest." As of today, the conventional oil "proven" reserves worldwide are about 1.3 trillion barrels (some of the latest additions are questionable), sufficient to last about 41.6 years at the present consumption level. An analysis of the need to find new reserves is presented in the graph below. It shows two sets of curves. The first set, the solid lines, using the left-hand scale, shows the years of reserves available as time changes. The second set, the broken lines, uses the right-hand axis and shows the need to find new reserves, expressed as a fraction of today's reserves. Figure No. 12 assumes that consumption will continue growing at a conservative estimate of 2% per year (the historical rate is actually approximately 3.5% per year).

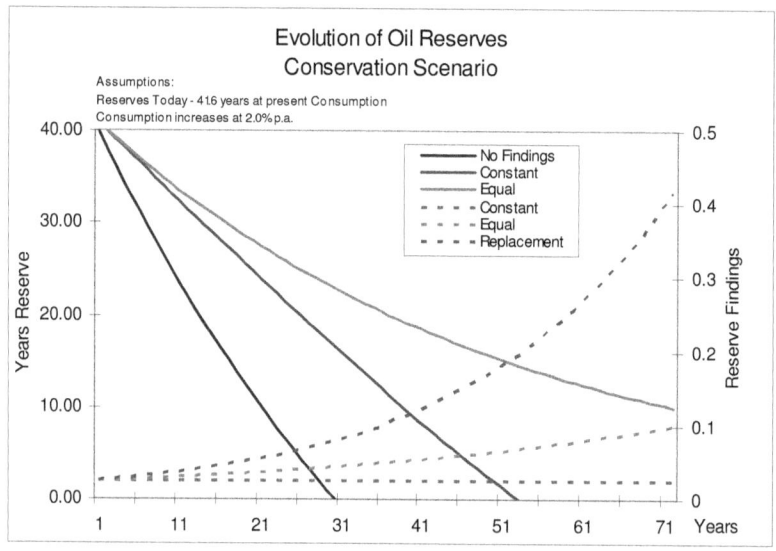

Figure 12 Evolution of Reserves

Assuming no new discoveries, the 41.6 years of reserves will only last about 29 years because reserves are measured at the current rate of consumption, without assuming new growth. If constant new finds were made for the next 30 - 40 years at the present level of consumption (i.e. 86 million barrels/day or about 31.4 billion barrels per year), the proven reserves would last only about 53 years, since new finds will be insufficient to replace the growing consumption. If the finds each year were equal to the consumption of the previous year, the reserves would last about one hundred years, because there would be a growing gap between the finds and consumption. If we were to replace the reserves by the amount consumed (i.e. maintain a level of proven reserves at 1.3 trillion barrels), the world would need to find every year a growing amount of oil (starting with about 31.4 billion barrels next year, and then increasing to about 38.2 billion barrels by the 10th year, 46.6 billion barrels in year 20, and 56.9 billion barrels in year 30), finding during the next 30 years basically the same amount of oil as we have today as proven reserves. If, on the other hand, we wanted to always maintain a level of reserves of 41.6 years, the world would need to find 44.2 billion barrels by year 10, 65.7 billion barrels by year 20 and 97.6 billion barrels by year 30, requiring the discovery of 1.750 trillion new barrels of oil in the next 30 years (which is about 40% more than today's proven reserves – excluding the new adjustments!).

In contrast to the figures given above, in the last 10 years, despite technological advances in exploration and recovery, coupled with the urgent need of the oil companies to discover new sources to maintain their economic viability and benefit from the current high prices, the world only found 47 billion barrels (less than two years of current consumption), and accounting gimmicks, mainly by Venezuela and Iran, increased reserves by 331 billion barrels.

In summary, while there is still oil in the ground, the possibility of finding the quantity of reserves (1.3 – 1.6 trillion barrels) in the next 30 years needed just to maintain the level of reserves at the present level, while maintaining production at current levels or even growing levels, is highly questionable.

Peaking Point

When a new oil field is discovered, oil usually initially flows un-aided, but as the pressure in the field decreases, pumping water or injecting steam (among other methods) is needed to recover additional oil from the field.

In 1956, M. King Hubbert, a well-known geologist at Shell, noticed that the production pattern of many oil wells tended to follow the typical probabilistic bell-shaped curve when production is plotted against time, with early production increasing quickly, as additional resources are brought in, and then starting to decline, irrespective of the methods used for the recovery of oil.

He suggested that production of oil fields follows the pattern of oil wells and also that the production level of different countries should follow the same pattern. Using this model, he predicted that production of oil in the lower 48 states would peak in 1971.

Hubbert's prediction was ridiculed by many. Oil production in the US was still rising then. However, production in the US peaked one year earlier than Hubbert had estimated, and since 1970, it has declined inexorably despite billions of dollars spent for exploration and the introduction of increasingly more expensive and efficient recovery methods.

Shale oil was known in Hubbert's times, but he never took it into consideration. Extraction of unconventional shale oil has increased drastically in the last decade, as new fracking wells have been brought into production, and in 2014, US production surpassed imports for the first time in more than 30 years.

Applying Hubbert's technique to world production, the "pessimistic" group believes that world oil production will peak before the end of the decade, if it has not done so already. After that, worldwide production of oil will decline.

Non-conventional oil

There are three large reserves of non-conventional oil:

1. Tar oil: Canada has the largest reserve in the world, reportedly about 1.7 trillion barrels trapped as tar oil (bitumen), of which about 10% is expected to be recoverable, mostly near the surface and susceptible to strip mining. Canada started extracting oil from the tar sands in the 80s and is now producing more oil from tar sands than from conventional sources. Extracting oil from the tar sands requires massive amounts of water (about five tons of water per ton of oil extracted) and energy (a ratio of energy recovered on energy invested or "EROEI" between five and six). Furthermore, it has a worrisome environmental impact, since the leftover sand sludge contains dissolved minerals that might seep out from the tailings ponds.

2. Shale oil: The US is endowed with a couple of large reservoirs of shale oil: the Green River basin (in Colorado, Montana and Utah), assumed to contain more than two trillion barrels of shale oil (kerogen), of which 6-10% might be recoverable, and the Bakken region, currently producing about 1.5 mbpd. There are also small reserves being exploited in Estonia, Russia and Brazil. The latter reserves are near the surface and are used almost as low quality coal, while the U.S. reserves are much deeper and require extraction with a high pressure mixture of water, sand and undisclosed chemicals. The EROEI for the American shale oil is expected to be even lower than for the Canadian tar oil.

3. Orimulsion: Venezuela has 270 billion barrels classified as reserves of a very viscous bitumen and very heavy crude oil in its Orinoco basin. This bitumen oil flows, albeit slowly and is being marketed in suspension form as Orimulsion, which can be burned directly

in a boiler. Its main drawback is its high sulfur (and other contaminants) content.

While the existence of non-conventional oil has been known for many years, the high processing cost and large requirements of water and energy requires a price of oil above $100/barrel to be attractive. Non-conventional oil requires more energy to extract or to produce synthetic oil, and thus, the return on energy is lower than that of conventional oil. Similarly, from the CO_2 emission point of view, non-conventional oil releases more CO_2 than conventional oil, because it takes energy to produce them.

The high prices at the end of the decade provided incentives to extract known non-conventional oil. Canada is now producing almost 3.75 million barrels per day from the tar sands, and production in the US from shale oil reached five million barrels per day in May 2015, with US total oil production exceeding imports for the first time in almost 40 years.

Natural Gas

Natural gas associated with oil was viewed as a nuisance prior to the 1950s, and was usually flared (burned) to dispose of it. A dry well (yielding only gas) was considered bad luck. Although there is the potential for gas to replace gasoline, most of the available gas has been committed to producing electricity and heating homes. Gas demand in the US is expected to double by 2030, with over 40 percent of the total demand used to generate electricity, as the vast majority of new electricity plants coming on line are powered by gas.

Total worldwide endowment is calculated to be between 6,000 and 7,000 trillion cubic feet. The US has doubled the amount of its reserves with fracking, but still has only less than 5% of the remaining global reserves and has used about 40% of its reserves. As with oil, most of the reserves are in Russia and the Middle East. Russia and Iran alone have 45% of the available reserves.

Gas field depletion is different from oil depletion. Once the pressure of the field is gone, there are no practical recovery mechanisms. A gas field yield usually 70-80% of the gas over its life-time. Producers seek to control the outflow to attain long-term, sustainable production.

In the US, the situation is similar to oil. After a period of stable reserves, starting in 2007, fracking produced a glut of natural gas, and the industry is pressing for export licenses.

Coal

Coal is a dirty fuel, emitting more CO_2 than natural gas, plus a lot of other nasty substances, such as sulfur, cadmium and mercury, some into the atmosphere and the rest as coal ash that is kept in ponds. However, coal is available in many places, it is inexpensive and the thermoelectric plants that burn it are mature and steady. The US has reserves for more than 200 years, and electricity generation with coal still accounts for more than 40% of the total. Both China and India have plenty of coal and are therefore building coal-based thermoelectric plants to satisfy their growing demand for electricity.

Prophets of Doom

So far, all the prophets of doom have been wrong, and I do not see any major benefit in adding my name to that list. Many optimistic predictions have also proven wrong, and I am not planning to add my name to that list either.

There are plenty of optimists that claim that the drop of oil prices appears to be the result of American ingenuity in extracting oil from rocks!

Sheikh Zaki Yamani, a Saudi Arabian who served as his country's oil minister three decades ago, once said, *"The Stone Age did not end for lack of stone, and the Oil Age will end long before the world runs out of oil."* The quote can be interpreted many ways: there is more oil to be found; new sources of energy that would be displacing oil are around the corner; or the cost of extracting some reserves is going to be very high.

The last interpretation is probably the right one. With increasing demand and limited resources, the price is likely to become unaffordable to many, dampening demand to the point that other alternatives, maybe unknown today, will be cheaper than oil. Oil at $100/barrel is abundant, but demand at even $60/barrel is limited. Gasoline sales in the US peaked in August 2005[7] at 399 million gallons per day (mgd) and bottomed in January 2014 at 322 million mgd. On a yearly basis (to avoid seasonal variations), the high price of gasoline has evaporated about 8.4% of the demand (about 32 million mgd) between 2005 and 2014. In other countries, the reduction has been more sanguine.

Low oil prices are as bad as high. Many producing countries need the revenue to continue subsidizing their people, and the drop in revenue translates into political unrest. Privately owned companies, large or small, need also high prices to be able to pay bank loans. With low prices, they cannot borrow more to drill other wells, so eventually production drops. If the demand for gasoline has dropped nowadays by 8%, prices need to go down even further to increase demand, which unfortunately translates into many possible bankruptcies. Even when there is oil in the ground, we are in for a rough ride.

My position is simple. I recognize that the unmatched energy density and versatility of fossil fuels have produced tremendous advances and benefits. However, given the possibility that US production of oil and natural gas might decline in the near future, I consider that the prudent thing to do is to assess what other alternatives are available, rather than foolishly use the limited oil we have to extract a little bit more from rocks. Instead we should be leaving some oil in the ground for those uses for which there are no quick substitutes.

The dual impact of peaking oil and concerns about CO_2 emissions or climate change makes it imperative that we immediately begin seriously exploration of all carbonless alternatives.

---

[7] US Energy Information Agency – US Gasoline Retail sales

Chapter 5 Electricity Generation

Electricity is not a primary source of energy (except for hydro-electricity). It is produced mainly by burning coal or natural gas, or by nuclear energy. Yet, electricity is the most adaptable form of energy. It can be used to power miniscule appliances, like an alarm clock or electric toothbrushes, or to move heavy loads, such as elevators, trains or trolleys. It is always available just by flipping a switch; it is noiseless, odorless and safe.

From its early beginnings as a laboratory curiosity in the second half of the 19th century, electricity has risen to become the most practical source of energy for both the home and the work place. The first generation plant selling electricity to customers was Pearl Street in New York capable of powering four hundred light bulbs (some 20 kWh). Electricity is now available almost everywhere, lighting homes and powering industries.

From the user's point of view today, nothing could be simpler. It is available everywhere (well, sometimes an extension cord could be needed!), it is always available (you just have to turn on the switch) and it can be used in a myriad of applications. We go to the store and buy the latest appliance or gadget, come home, plug it in and it works. We do not have to worry about overloading the system or having to disconnect other gadgets.

Demand for electricity worldwide has increased steadily at about a 3.5% rate, reaching 22.8 PWh (22.8×10^{15} Wh). That is about 3,000 kWh/y per capita worldwide. The US demand has also grown, albeit at a much slower rate because penetration is almost complete. Table No. 4 shows worldwide and US demand from 2000 to 2012.

From its humble origins, three characteristics of the supply of electricity were quickly recognized: (i) generation and distribution are capital intensive, requiring large investments for the generating equipment and distribution systems; (ii) it provided a natural monopoly because it does not make sense to construct multiple distribution systems, and; (iii) it needed to be monitored and regulated to prevent abuses from greedy owners seeking excessive profits. All over the world, electricity prices are either regulated or provided by state-owned companies.

Electricity Demand (TWh)			
	World	US	(%)
2000	14,627	3,802	26.0
2001	14,879	3,737	25.1
2002	15,393	3,858	25.1
2003	15,927	3,883	24.4
2004	16,692	3,971	23.8
2005	17,330	4,055	23.4
2006	18,033	4,065	22.5
2007	18,067	4,157	23.0
2008	19,157	4,119	21.5
2009	19,093	3,950	20.7
2010	20,437	4,125	20.2
2011	21,182	4,100	19.4
2012	21,532	4,048	18.8

Table 4 World and US Electricity Demand

With occasional hiccups, the system has evolved to provide an extremely reliable service at very reasonable prices. Initially, demand was limited to lighting, but the introduction of trolleys to many cities brought diurnal demand for electricity. The introduction of electricity provided business opportunities to many entrepreneurs to provide electricity to small cities or towns. As demand grew, there was massive consolidation into larger utilities, but there are still many small private or municipally owned companies. Transmission lines were added to bring electricity from remote hydroelectric generators to cities and then to interconnect cities to improve reliability.

In many places, the system is open – anyone can generate electricity and sell it to their neighbors, but economies of scale have favored integration into large companies that generate their own electricity and sell to millions of customers within a well-defined distribution area. These large companies have strong political and market clout, but overall the open system and the existence of regulatory bodies have worked well.

A possible projection worldwide of the supply of energy for the next 30 years, assuming business as usual, broken down by source, is shown in Figure No. 13.

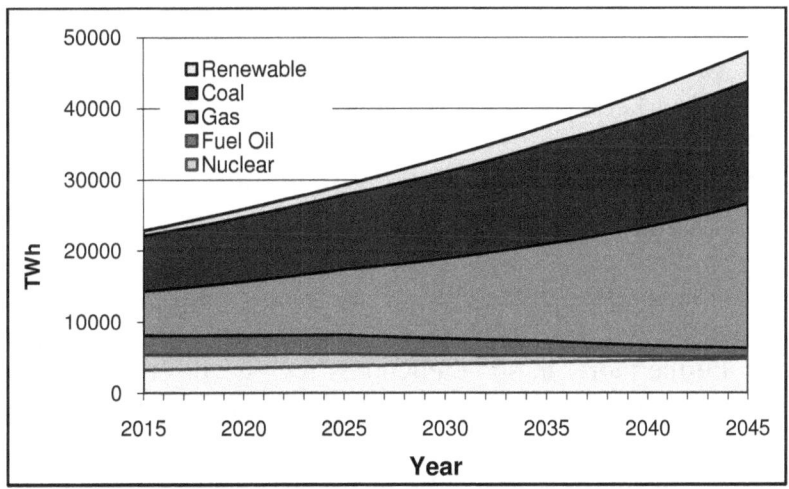

Figure 13 Possible Future Electricity Generation

Figure No. 14 depicts one of many scenarios. The graph shows a modest increase in hydroelectricity, since most of the resources have been tapped. Nuclear energy is expected to decrease over-time (no new plants in the US are being constructed). Natural gas shows a big increase because it is considered a cleaner alternative. Fuel oil is also expected to decline, leaving coal still as a main source of energy. There is also an increasing, but slow penetration of renewable energies. The drastic rise of natural gas as the fuel of choice is at times tempered by the volatility of natural gas prices, but there are still many plants in the planning stage. In this scenario, renewable energy will represent 8.2% of the total in the year 2045 and might be doable by a market-oriented effort without political impetus.

Other scenarios can be constructed. Nuclear energy proponents insist that it does not produce green house gasses and thus should be the alternative of choice; that there have been more accidents and deaths from thermoelectric plants or refineries than from nuclear plants in the US; that the Chernobyl accident was the product of people disconnecting security backup systems and that contamination escaped because the plant did not have a containment building unlike American plants, and; finally that getting all the electricity needed in a person's lifetime would produce only about two pounds of nuclear waste, while burning coal would emit about 13,000 pounds per year of CO_2 per person.

Transmission, Generation and Distribution

Many countries have divided the system into its three main components: transmission, generation and distribution. Each component is remunerated differently. They usually limit the ownership by a person or group to only one kind of asset to spur competition. Furthermore, in some countries they have forced distribution companies to open their system so that other companies can sell electricity within their coverage area, paying "rent" for the use of the distribution facilities.

Transmission is a natural monopoly – the assets are placed in service to connect generators with distribution companies. The value of the assets is known and it is easy to devise a return-on-assets formula. The owner of the transmission line gets paid the same amount irrespective of fluctuations in demand. The clients are usually few, big, sophisticated companies and the allocation is straightforward. Some segments of a transmission line might have only two clients – one generator and one distributor. With integration of systems, transmission lines improve overall system reliability, allowing electricity to flow from resource-rich areas to resource-constrained ones, minimizing disruptions due to equipment failures or localized surges in demand.

Distribution is also another natural monopoly, but with many clients with different consumption levels. One client might be an industry demanding several megawatts and another client a small residence consuming little electricity. Distribution companies lay thousands of miles of lines connecting almost everybody. Distribution is usually regulated to assure uniform tariffs within the area. The regulations define the level of profitability for the distribution company by setting tariffs that will cover all expenses and the targeted profit. The profitability depends on a forecast of consumption, expenses, technical and commercial losses, billing and collection and even the frequency of reading the meter for billing purposes. Regulators define the allowable level of losses to require distributing companies to have an efficient system and thus limit the losses that can be passed on to the customers. Since there are often differences between real and forecasted figures, there is a continuous tug of war between regulators and distribution companies. Setting the tariffs is complicated by political consequences. Supplying electricity to industries at a higher voltage is more efficient than supplying electricity to residential users at a lower voltage, who are more spread out and thus require a bigger, costlier distribution system. It is cheaper to sell electricity to large clients, but there are many more small consumers who can use their far greater numbers to influence elections. Traditionally, there is a small cross subsidy from large consumers to small consumers.

Generation is usually open – anyone interested in generating electricity can construct and operate a generation plant (after meeting municipal and state requirements) and be connected via transmission lines to consuming centers, selling electricity to distribution companies or to private clients. To prevent unfair competition or the transferring of profits between regulated and unregulated sectors, distribution companies are often, but not always, prevented from owning generating assets.

The cost of generating electricity has a tremendous repercussion on tariffs. With economies of scale, generation equipment tends to be large, and their owners sell electricity to distribution companies. Almost universally nowadays, the cost of electricity is a pass-through, allowing distribution companies to recover the real cost of purchasing electricity, which is paid by the customer.

Two characteristics of the electric sector prevent the implementation of simple solutions. The first one is storage. It is difficult to store electricity. Energy has to be stored in a form that can be later converted into electricity – a battery, pumping water or fly wheels, but there are penalties (losses) in doing so. Storage in batteries or fly wheels is expensive, and pumping water is limited to sites that have the proper topography. The second is fluctuating demand. There is a continuous variation in the demand for electricity. It changes minute to minute during the day and seasonally. Demand at 2 a.m. in residential areas is only a fraction of the demand during early evening hours. Demand in spring is only a fraction of that on a hot summer day.

Fluctuating Demand

Figure No. 14 shows a typical consumption pattern in an upscale house during a late summer day on a week end. If the day had been hotter, the air conditioner would have been working more and additional peaks of demand would have appeared earlier. If it had been a work day, with the kids at school and the parents working, the early afternoon peak would not have materialized. While the average consumption in the graph is 3 kWh during the day (horizontal line), peak demand reaches 14 kW during those times when the air conditioner, water heater, microwave, refrigerator, stove, TV and other appliances are on simultaneously. The blips during the late evening and early morning are produced by refrigerators - freezers that kick in sporadically. The curved line shows typical output of a PV panel.

Fortunately, with thousand of houses, the curves are smoothed out. The following set of graphs (Figure No. 15) shows some typical patterns, with the caveat that there is no such thing as a standard day or even sector. There are hot days and nice days in the summer and cold and nice days in the winter. Commercial consumption includes a variety of loads, each with its own characteristics. The load from an office building differs from the load of a shopping mall or hotel or movie theater. Industry

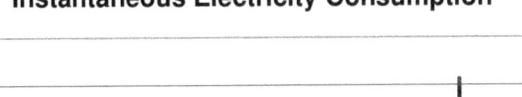

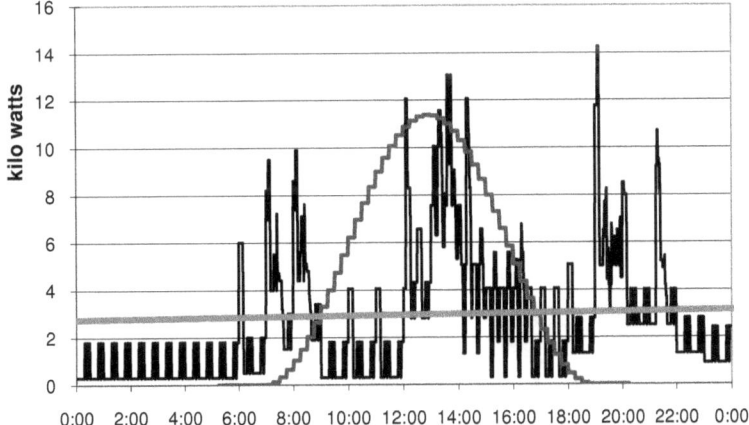

Figure 14 Instantaneous Electricity Consumption

encompasses a variety of operations, some working 24/7, while others only work one shift. The graphs assume an even distribution between industrial, commercial and residential demand, which resembles the national average, but is not applicable to North Carolina or Washington DC, which have little industrial demand, and might be off for Pennsylvania or New Jersey, with their larger industrial demand.

The short-term peaks and valleys disappear, but the general shape of the graph clearly shows noticeable peaks depending on the season, an early morning peak, one around noon and a longer evening peak. During the year, the above graph presents variations. During the winter, for a house heated with a heat pump, there will be many more high demand blips during the evening and night, as the heat pump kicks in to maintain the house at the desired temperature, with less frequent blips during the day, as the sun provides some heat. During the summer months, the blips will be more frequent during the day, as the heat pump tries to maintain the house at the desired temperature, with lesser demand during the night.

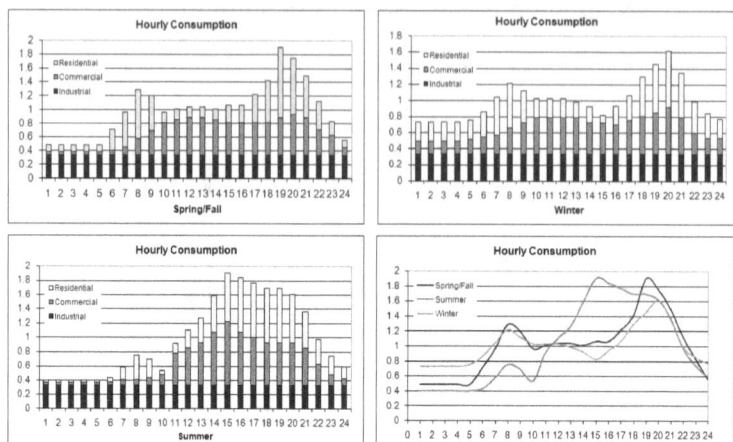

Figure 15 Example of Demand

The commercial demand for electricity presents a steadier level of consumption during business hours, as offices and retail stores open and provide light and cooling/heating for their occupants. During spring and fall, consumption during the day is rather constant and substantially higher than during non-business hours. During the summer and winter, the heating/cooling requirements become the bulk demand for electricity.

Demand for electricity by heavy industry shows less variation during the day and over seasons if the plants operate continuously, but for light industry working only one shift, the graph would be similar to the one for commercial demand.

The utilities have done a remarkable job in providing steady service, despite large daily fluctuations. Typically, in the summer months, demand at night can be 40% of the average while during the early afternoon hours it can be almost 200% of the average. It is almost a miracle that we do not have to reset the clocks on our appliances more than once or twice a year.

Generation

Public utilities generate electricity mostly with large, efficient thermoelectric plants. The source of heat or thermal energy might be nuclear power, coal, natural gas or heavy fuels like fuel oil or even diesel. Coal is the predominant source of energy in the U.S., generating about 49% of all electricity, followed by natural gas at 20% and nuclear energy contributing 19.4%, all together accounting for 88.4% of all generation. The remainder is hydroelectricity (about 7%), renewable (about 2.4%) and about 2.2% from heavy fuels or other products.

Given the fluctuating demand of electricity and the difficulty in storing it, generation is broken down into:

1. base load plants that operate all the time, usually at peak capacity, and are usually the lowest cost generators;
2. cycling plants that also operate most of the time, but can be throttled up and down, capable of operating at reasonable efficiencies over a wide range of loads and/or be maintained warm, in standby mode at reasonable cost;
3. peaking plants, units that can be turned on and off with relative ease to meet changing demand.

Nuclear plants are usually base load plants, thermoelectric combined cycle and coal generation are employed in cycling plants, while gas turbines and hydroelectricity (when available) are typically used in peaking plants. Centrally located dispatch centers monitor the lines all the time and instruct generators to throttle up or bring additional equipment on line if the frequency (electricity is delivered at 60 cycles per second) starts lagging or to throttle down or remove equipment if the frequency increases. To be able to cover equipment malfunctions, dispatch centers maintain equipment warm, in reserve or on standby, to be ready to come on line within seconds/minutes. Failure to bring additional equipment on line results in brownouts and eventually disconnection of some circuits.

To increase the robustness of the system, many utilities are interconnected by transmission lines. Failure of a 200 MW plant in a 2,000 MW system would trigger the disconnection of many circuits, but when the system is interconnected to a 5,000 or 10,000 MW systems, the effects are much less noticeable.

Table No. 5 shows a comparison of the costs of generating electricity utilizing both conventional systems and alternative sources, based on ballpark figures of the investment required, the price of natural gas at $4/million BTU and a $3/million BTU cost for biomass and coal, a 20-year life and no O&M (operation and maintenance) charges.

Cost of Generating Electricity						
Source	Efficiency	Cost of Fuel ($/MWh)	Investment ($/kw)	Availability (%)	Capital Cost ($/MWh)	Monomic Cost ($/MWh)
Gas Turbine (peaking)	34%	40.15	500	30	9.51	49.67
Thermoelectric - coal/steam	47%	21.79	800	95	4.81	26.59
Combined Cycle	56%	24.38	1,000	95	6.01	30.39
Hydro - run of the river			600	60	5.71	5.71
Hydro - dam for storage			1,200	90	7.61	7.61
Wind Turbines			1,000	25	22.83	22.83
Photovoltaic			2,500	16	89.18	89.18
Solar Thermal High Temp.	32%		7,600	41	105.80	105.80
Solar Thermal Low Temp. *	16%		4,000	25	91.32	91.32
Biomass	28%	36.57	1,500	75	11.42	47.98

Table 5 Comparison of Generating Costs

If the price of natural gas continues in the range of $2.5 to $3.5/MBTU, renewable energies will have a tough time competing. In the past, the US enjoyed prices of 5¢/kWh ($50/MWh) because the historical price of natural gas was about $2/million BTU. The intermittent nature of renewable energy is an unwelcome nuisance or complication to the system. Even when combined cycle plants can be throttled up and down, there are limits to the flexibility and there are delays in response due to lag times. If the natural gas price were to double, just like in Europe, or to triple, as the current prices have in Japan, the price of electricity in the US would double or triple to reflect the fuel cost.

The Electric System in the US

Electricity generation in the US[8] was about 4.1 PWhr/y (4×10^{15} watts) in 2014. Total generation capacity is about 986,000 MW. There are more than 125 million residential customers consuming, on average, 10,600 kwh per year. Consumption is evenly divided between residential, commercial and industrial. The system is huge – there are a couple of hundred thousand miles of transmission lines and millions of miles of distribution lines. Total investment of the electric sector in generation, transmission and distribution at book value is likely to exceed $3 trillion.

The electric system has evolved over time into a robust, redundantly interconnected system capable of withstanding equipment failures or broken lines, with disruptions affecting only few customers. The integration into large companies allows economies of scale and synergies producing a very reliable service at reasonable prices.

Selling electricity is a complex business. Utilities are many times caught between a rock and a hard place. Because the business is regulated, they are required to be cost conscious with their investments, yet they have to provide an extremely reliable service. Investment in quality/reliability is at times characterized by regulators as "gold plating" the system to justify higher tariffs. The size of crews to be able to respond quickly to system failures or line maintenance is usually questioned, and then there are penalties imposed for system interruptions. Utilities executives are continuously fighting the regulators for cost reductions everywhere and then have to listen to politicians insisting on renewable generation, even if it is at a higher cost. If the utilities succeed in cutting costs one year or tariff period, the regulators might take that year or period as the new efficiency yardstick.

Given a tariff structure, a higher demand for electricity produces higher profits for the utilities, while a decrease in demand lowers

---

[8] Energy Information Agency – data for 2014.

profitability. It is therefore not surprising that the utilities do not recommend conservation.

The constant tug of war between regulators and utilities has produced a risk averse management sector, reactive rather than proactive to the renewable energy challenges. Installing intermittent generation requires throttling down and lower efficiencies of base and combined cycle plants, opening a new front in the continuous struggle between utilities and regulators.

Utilities executives, although they claim to be interested in renewable energy, are tacitly blocking its proliferation by insisting that the feed-in-tariff should be lower than the tariff paid for a newly constructed combined cycle (the most efficient) thermoelectric plant, because of the intermittent nature of the renewable energy generation, and point out that they will be willing to pay such a tariff if the dispatch-ability of the system is similar to that of a thermoelectric plant.

Although the penetration of renewable energy is still low, utilities are worried that massive renewable energy installations will complicate the balancing of the system that would have to accommodate the diurnal generation of PV panels and the variability inherent in wind energy, and that, at the end of the day, the utilities are the ones who will be penalized if the system cannot adapt quickly enough to passing clouds or wind gusts.

The public in general has stated its preference for renewable energy. However, if given the option of more expensive and less reliable renewable energy versus continuing with the current reliable and cheaper energy we have, the overwhelming majority will choose the continuation of current practices, with reliable and less expensive service.

Energy Storage
A nasty complication of electricity is that it is difficult to store. An inexpensive solution for storing energy is needed or solar energy will continue to be only a diurnal source of electricity.

The amount of energy that can be stored is proportional to the mass of the storing substance. Fossil fuels have a compelling advantage regarding energy storage density. As a point of comparison, the storage density of gasoline is about 46 MJ/kg.

A quick analysis of available energy storage possibilities is presented in Table No. 6. The list does not include all possibilities. Flywheels or ultra capacitors can provide instantaneous energy, but not for several hours a day. The investment cost is a guessed estimate based on an installation to store 6 MWh. In the case of pumping water, the investment would depend on the topography; for compressed air on the existence of a suitable cavern; and for salts, it includes the cost of heat exchangers, pumps and salts. The life span of the installations is also estimated. The short time span for salts assumes there will be some corrosion on the equipment. While there might be some economies of scale, size drastically increases the level of complexity.

Storage Cost					
	Investment ($'000)	Round Trip Efficiency (%)	Life span (years)	Cost ($/MWh)	Storage Density (kJ/kg)
Water Pumping	1,000	75	20	22.83	1 @ 100 m
Compressed Air Energy Storage	2,500	60	20	57.08	~ 50
Batteries	2,500	85	5	228.31	Pb ~ 140 Li ~ 500
Hot Water	750	80	25	13.70	125
Thermoil	3,000	85	25	54.79	65
Molten Salts (wide)	2,500	85	15	76.10	200
Molten Salts (narrow)	3,000	85	12	114.16	360

Table 6 Cost Comparison of Storage [9]

Pumping water is very appealing, but unfortunately is limited by the topography. There are no hills to pump water uphill in Kansas City. The investment cost of pumping water varies with the topography, which will define where the two reservoirs will be located. The main problem with water pumping is the low energy density. To produce 6 MWh, with a 100 m rise, the upper reservoir needs to hold 21,600 m^3 (18 acre-feet). Assuming free

[9] simple calculations based on described assumptions, bulk part storage densities and life expectations -

(or inexpensive) energy six hours a day, a pipe 36 inches in diameter is needed to pump the water uphill. The round trip efficiency should be about 75% (pump efficiency plus water evaporation and permeability). The capital cost per MWh was calculated by dividing the estimated investment by the energy stored (one MWh times six hours times the life of the system). If the same installation could be used for eight hours a day, the cost would drop to $17.12/MWh. The actual cost will depend on the topography of the installation, the numbers of hours you have free or inexpensive energy and the actual investment.

Compressed air energy storage (CAES) is another possibility. To be able to store energy, a large, impermeable cavern is needed to store the air. Unfortunately it is also limited by the availability of caverns, salt deposits, abandoned mines, aquifers or spent oil or gas fields. The second problem is the need for natural gas to achieve reasonable round trip efficiencies. When air is compressed, it heats up, but when compressed air expands, it cools down. While the heat rate of a gas turbine using compressed air and natural gas might be attractive, fuel is still needed to generate renewable energy on demand. The round trip efficiency is about 60%. The shown cost estimate is more speculative. Needed is a three or four-stage compressor, large areas of radiators, a suitable cavern and a modified gas turbine with a pre-burner to accept the compressed air. Only two installations exist: McIntosh in Alabama and one in Germany.

The third option is storing electricity in batteries. It has been well researched for more than a century by large companies and universities, and while some improvements have been made, the life span of batteries is limited. The main problems are cost and short life (5-7 years) because the potential storage degrades over time. The cost was estimated based on lead-acid $250 100 A/h deep discharge cycle batteries. Lithium, sodium-sulfur and other flow batteries, while potentially cheaper, are still not being produced in large enough quantities to achieve mass production costs. If and when better batteries become available, they could be the solution.

Another option is thermal storage. There are basically two choices – phase change heat and latent heat. The first alternative has greater storage density. Phase change from liquid to vapor increases the volume drastically and becomes impractical. Phase change from solid to liquid has a complicated heat transfer mechanism in the solid-liquid interface and the additional expense of the material. Ideally, it should be operated at the melting point, but then the heat transfer mechanism is limited to conduction by a solid. Increasing the operating temperature range to have a liquid phase interface to improve heat transfer dilutes the overall heat capacity of the melting substance.

Water has a heat capacity of 4.18 kJ/kg °C. If water is used within a 30 °C range, it can store 125 kJ/kg which compares well with other storage media. However, the Carnot efficiency for an engine operating 30°C above ambient temperature is not as large as an engine operating at high temperatures.

The volume necessary to store energy grows quickly with the amount to be stored.

Substance	Heat Capacity kJ/kg	Radius of Sphere (m)		
		6 MWh	60 MWh	600 MWh
Thermoil	65	7.68	3.55	35.66
Water	125	6.17	13.29	28.64
Molten Salts (wide)	200	5.28	11.38	24.51
Molten Salts (narrow)	300	4.34	9.35	20.15

Table 7 Needed Radius of Sphere for Storage

Table No. 7 shows the radius of the sphere required for the different substances to store enough energy to produce 1, 10 and 100 MWh for six hours with a mid-temperature thermal engine with 17.5% efficiency. None of the spheres can be transported assembled. They will have to be transported in multiple segments and welded on site. A cube, formed with flat steel might be easier to transport in segments, but still have to be welded.

Using a more efficient thermal engine does not reduce the dimensions much. The radius of the sphere would be 20% smaller for a 30% efficient thermal engine. Beautiful as it sounds, unfortunately we are not yet out of the woods. To store 1 MWh_{th} with thermal salts storing 360 kJ/kg, we would need about 10 tons. To convert it into 1 MWh_e, we would need 30 tons or more. The US consumed on average 11 million MWh per day, we would need 330 million tons of salts to store enough energy for a single day.

The conclusion that the market, utilities and research laboratories have reached is correct: it is expensive, massive and does not mesh easily for utility-size installations. However, thermal storage can work for distributed generation.

Based on this analysis, the molten salts alternative can be quickly discarded because it adds complexity without a clear economic advantage. We can also discard the thermoil alternative, which offers the possibility of a low pressure storage tank, because of the price of thermoil. For a system to generate 250 kWh for six hours, the volume required for thermoil and pressurized hot water are 475 and 246 cubic meters, respectively. Water might cost $1/m^3$, but thermoil is about $1,000/m^3$. The cost difference is about $490,000. By elimination, we are left with a pressurized hot water storage tank as the best alternative. Water is inexpensive, simple to use and could easily have a 25 year lifespan. The main problem is that its vapor pressure increases exponentially with temperature.

The efficiency of a thermal engine is a function of the difference in temperature between the hot fluid and the cold sinks. A typical thermoelectric engine operating at steam's critical temperature is about 33% efficient and requires steam at about 700 psi. The thickness of the wall of the storage tank is a function of the pressure and the diameter of the sphere. A steel sphere holding $250 m^3$ of water at 700 psi would require a wall thickness of 1 5/8 inches. If the pressure is 150 psi, the thickness of the wall needed is only 5/16 of an inch. Another possibility is a concrete tank. Unfortunately concrete works well under compression, but not very well under tension, which means that the concrete would need to be reinforced with steel and would have thick walls. But concrete is much cheaper than steel.

Another possibility is storing energy with thermo-chemical reactions, reversible reactions whose equilibrium is sensitive to temperature changes. It includes the decomposition of metal hydroxides, oxides, peroxides, ammoniated salts or carbonates. The amount stored is a function of the amount of reactants, the heat of the reaction and the fraction of matter reacting.

Usually the products and the reactants must be separated to avoid a spontaneous reversible reaction if the temperature drops. Many times, the products are valuable chemical products that must be cooled, separated and stored for further use to regenerate the substance. Some organic materials decompose over time into a variety of products, which means the spent reactant needs to be replaced periodically. Some reactions require applying high pressure to regenerate the product.

The final alternative is hydrogen. It is complicated, expensive and imposes a stiff efficiency penalty. It will be discussed later in the Renewable Energy section.

The conclusion reached is that the best way to store energy is in the form of thermal storage for a small system operating at mid temperatures (about 180°C.) The size of the tank is still a challenge. Given the heights of overpasses and lengths of platforms, the maximum size of a tank that can be transported is about 12 ft in diameter and 50 ft in length, capable of holding about 60 m^3. To store enough energy to produce 250 kWh for six hours would require four of these tanks or four segments welded at the site to make a single tank.

Thermal storage is a viable alternative, albeit with a big complication: thermal storage degrades quickly. While a battery might hold its charge for several weeks or months while not in use, the temperature of a heated material drops quickly despite good insulation. A 3-5 day period without sun light could mean a week without electricity until the temperature of the tank could be raised to the desired or needed operating temperature.

If it were not for this little detail of CO_2 emissions, the problem of energy storage could be easily solved with fossil fuels (a liter of gasoline stores about the same amount of energy as the latent heat of 100 kg of eutectic salts). You could have a large exposed pile of coal, store an oil product in a tank or natural gas injected into spent natural gas fields and use it whenever it is needed.

To overcome the lack of storage, the electric utilities have come up with the grandiose plan of a smart grid, which is a way of shedding loads based on the instantaneous price of energy. The idea sounds great on paper. You put a smart meter in every house and let the customers set their threshold price. The frugal, penny pincher or cost conscious consumers would use electricity when the price is less than their threshold. For some uses, the consumer might be prepared to pay a higher price, but for other uses (i.e. washing clothes) the need can postponed until the price of electricity is lower. The appliances would need to talk to the smart meter to consume electricity only when the price is below the threshold. The utilities would need to broadcast the instantaneous price of electricity. The management of the public utilities believes that people would very sensitive to fluctuations and would not watch TV or use their air conditioning until the price goes down so that they can save one full dollar a day. I am afraid that the smart grid is wishful thinking and is bound to fail because electricity demand is quite inelastic.

Chapter 6 Solar Energy

In this chapter, I am going to be talking about the sun as a source of energy. We, plants and animals, owe our lives to the Sun, but as a resource to provide the energy to maintain the standard of living our civilization has today, it is too diluted and variable.

The purpose of this chapter is to provide a frank assessment of the availability of solar energy. While some state that the Earth receives daily thousands of times the amount of energy needed by mankind in a year, the resource is very diluted and changes during the day and seasonally. That translates into requiring a huge area to capture meaningful amounts of energy.

The Sun is a sphere of extremely hot gaseous matter of 0.9 million miles (1.4 million km) diameter, located about 94 million miles (150 million km) away from the Earth. Although the Sun has an effective blackbody temperature of only 5777 °K, the temperature in the nuclei is estimated to be between 8 and 40 million °K. The Sun is in effect a continuous fusion reactor, contained by gravity force. The most important reaction is the combination of four hydrogen nuclei (one proton each) into a helium nucleus (two protons and two neutrons – a neutron is similar to a proton, but without an electric charge). The mass of the helium nucleus is less than the mass of the four hydrogen nuclei. The mass lost in the reaction is converted into energy.

The bulk of the conversion takes place near the Sun's center, where the temperature and density are highest. Moving away from the center, both the temperature and density decrease. At about 0.6 million miles from the Sun's center, the temperature has dropped to 100,000°K and the density is similar to our air density. At the outside, in a region called the photosphere, the temperature is about 5000°K and the density is 0.01% the density of air at sea level. The emitted solar radiation is really a composite result of several layers in the Sun that emit and absorb radiation at different wavelengths.

The solar constant G_{SC} is a measure of the energy received from the Sun per unit time on a unit area of surface perpendicular to the direction of propagation of the radiation at the mean Earth-Sun distance outside the atmosphere. The value of the solar constant is 1367 W/m^2, or 433 BTU/hr ft^2, or 4.92 MJ/m^2. The distance between the Sun and the Earth varies approximately 1.7% during the year, due to the Earth's orbital eccentricity. This leads to variations in the radiation flux in the range of ±3.3%, with lower values at the beginning of July.

Solar Irradiation

The solar radiation is a combination of radiation with wavelengths ranging from 250 nm to 8,000 nm, (the visible spectrum is between 380 nm and 750nm), with about 48% of the energy in the visible region. The spectrum of extraterrestrial solar radiation energy levels of different wavelengths is shown in Figure No. 16.

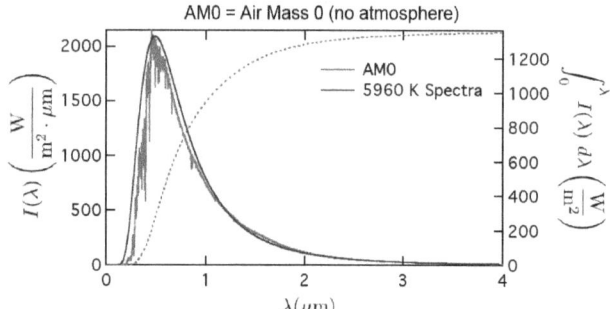

Figure 16 Solar Irradiation

As the radiation passes through our atmosphere, some of the energy is scattered as it collides with air molecules. Some of the energy is absorbed by water vapor, CO_2 and other gases present in the atmosphere.

When the Sun is at its zenith (directly overhead), the distance traveled by the rays through the air is the minimum. The path increases with the zenith angle of the Sun. The dilution of the solar irradiation caused by the atmosphere is shown in Figure No. 17.

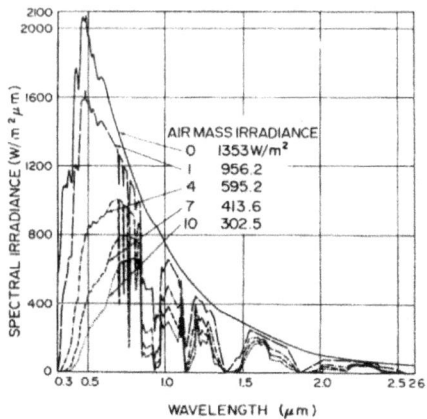

Figure 17 Irradiation on Earth

In addition to the attenuation of the solar radiation by passing through thicker air mass when the Sun is not in its zenith, the intensity of the radiation flux changes with the angle between the Sun and the surface, even if the surface is tilted. Since the Sun is moving continuously, the intensity of the solar radiation is also changing. The position of the Sun changes during the course of the day as it moves east to west and during the year as it appears to move north to south.

Figure No. 18 shows the Sun's position during the year and during the day as seen in San Francisco, California.

The graph shows that on December 21, the Sun will only raise above the horizon some 28.5°, and that on June 22, it will rise to 76°. The graph also shows that on March 21 or September 23, the Sun rises in the east and sets in the west, but that in December it rises about 30° south of east and sets about 30° south of west, while on June 22 rises and sets about 30° north of east and west respectively. The graph also depicts the length of the day, which is about 9.5 hours long on December 21 and about 14.25 hours long on June 22.

A similar graph can be constructed plotting solar irradiation, rather than the azimuth angle, allowing the visualization of the variation of solar irradiation through the day and the year.

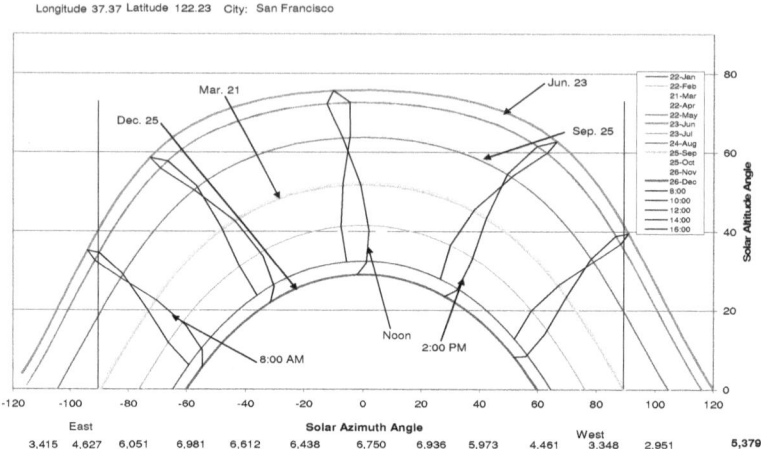

Figure 18 Sun's Position over San Francisco

Table No. 8, expressed in MJ/m^2, compares the data calculated against data reported in the literature. The variations can be explained by cloud cover, the effect of humidity or the difference between monthly average figures, as opposed to calculating numbers for one specific day in the month. However, for the purpose of this book, the calculated figures are sufficient to illustrate the wide variation of solar energy availability throughout the year.

Table No. 8 illustrates one of the main problems with solar energy. During the winter, solar irradiation is about 35-40% of the amount available in the summer, and therefore any solar energy system would is dimensioned to produce the needed energy in the winter, would result in much larger generation or underutilization in the summer. Can't win!

Calculated/Observed Irradiation (MJ/m² d)		
Date	Observed	Calculated
January 21	8.03	10.10
February 21	11.45	13.77
March 21	16.51	18.62
April 21	21.79	22.06
May 21	25.26	24.41
June 21	26.97	25.16
July 21	27.14	24.35
August 21	24.02	22.24
September 21	19.77	19.33
October 21	13.91	14.88
November 21	9.32	10.85
December 21	7.26	8.99
Average	17.62	17.90

Table 8 Solar Irradiation in San Francisco CA

Cloud cover has a substantial impact on the actual solar irradiation. Maps of the availability of solar irradiation show that the South west receives the largest amount of solar energy in the US, even though parts of Florida and Texas are further south.

Tilting Collectors

The simple act of tilting the collector modifies the energy that can be captured. Figure No. 19 below shows the energy received on a tilted collector located in Miami, Florida.

By tilting the collector or the photovoltaic panel facing south (in the northern hemisphere) at an angle similar to the location's latitude, the collector/photovoltaic panel is capable of collecting additional energy in the winter, while receiving energy at a lower angle during the summer with less time, because the collector cannot see the Sun while it is north of the east/west line.

There is no need for further elaboration. We all are familiar with the lovely weather in spring, the nice crisp days of October, the torrid heat of July and the freezing cold of January. We all know that in the summer, the days are long and in the winter, when we get up, it is still dark outside and when we leave work it is dark again. With a PV panel, you get more energy in the summer than in the winter (for the sake of the argument, in San Francisco, a PV panel would produce 2.5 times more energy in the summer).

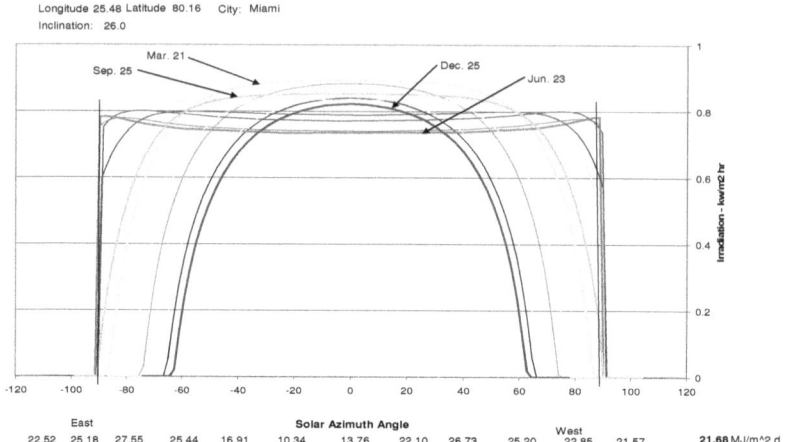

Figure 19 Solar Irradiation - Tilted Collector - Miami

Furthermore, although the available energy hitting an area on a cloudless summer day near the tropics can exceed 20 MJ/d m², the equivalent energy of about two cups of gasoline, we would be lucky if we could harness more than 20% of that energy. To capture the equivalent of a gallon of gasoline per day, we need about 40 m² of solar collectors (about 400 sqft) in the summer and some 100 m² or about 1,000 sqft in the winter. On average, one acre of land is capable of collecting the equivalent of about 60 gallons of gasoline per day. Solar energy is a diluted energy!

The final complication is cost. As you might imagine, it is costly. How are we going to ease our dependence on cheap, reliable fossil fuels with such a finicky resource? The consensus has been let's wait for some new technology to evolve. Fusion energy is around the corner and it will solve all our problems. We are just worrying about nothing.

Chapter 7 Renewable Energy and Limitations

In the previous chapters, despite the limitation of solar energy, I suggested that it would be foolish to believe oil will last for several generations and that planet Earth will quickly correct this abnormal CO_2 increase. Thus, I recommended that we seriously study all possible alternatives.

There are only three sources of energy: fossil fuels, nuclear energy and renewable energy (which run the gamut from hydro to burning dung for heat/cooking). These three sources cover four basic needs: electricity, transportation, heating and manufacturing (plastic, chemicals, glass, cement, steel, fertilizer, etc.).

Fossil fuels include natural gas, coal and oil. A large portion of the refined oil is used as fuel for transportation. Without oil, nearly all mechanized means of transportation would come to a standstill. Except for a few electric vehicles and even fewer demonstration fuel cell vehicles, the alternatives are horses, walking or bicycling. There are four potential avenues for transportation without oil: synthetic fuels, biofuels, electricity or hydrogen. The first three have problems. Synthetic fuels do not solve the CO_2 problem; as a matter of fact, they exacerbate it. Biofuels compete with food, and it is unlikely that they could replace all fossil fuels anyhow. Finally, hydrogen is not a source of energy; there are no deposits of hydrogen; it has to be separated from other compounds, which requires energy.

The U.S is endowed with plenty of coal and could continue producing electricity for many years using coal and natural gas. Unfortunately, burning fossil fuels exacerbates the green house gas emissions. There are several ways to produce electricity without fossil fuels: nuclear energy, wind energy, solar energy (either solar thermal or PV panels), biomass, geothermal, hydro and ocean (thermal, wave or tide) resources.

There is no single solution for the global problem. There are countries endowed with abundant solar resources, but little vegetation for biomass. There are countries with abundant land, water and manpower to implement a biomass/biofuel strategy. There are tropical islands that could use ocean thermal energy, and other places with abundant wind or geothermal resources. Each country/location would have to use the appropriate source for the region. Furthermore, there are places that might want or need to rely on nuclear energy.

The same applies to the US. Demand for energy in large cities like New York or San Francisco will prevent them from ever becoming energy independent – they do not have sufficient land to be able to capture significant amounts of solar or wind energy, and thus will have to purchase energy from neighboring counties or the state. New Hampshire has less solar irradiation than Nevada, Tennessee has little wind and Arizona, Nevada and New Mexico have little biomass.

Below is a discussion of each alternative and its limitations. The order does not reflect any particular preference or endorsement.

Nuclear Energy

Although nuclear energy is not a renewable energy, it is a clean energy source (at least with respect to greenhouse gasses!). It is the most controversial form of energy, mainly because of fears raised during the Cold War, when the threat of annihilation by nuclear war exaggerated the dangers of radiation exposure. There is massive mistrust and misinformation about the dangers of radiation. I am not saying that radiation is good, but we know enough about the dangers to have safe protocols and redundant backups to make it quite safe. Of course, there is always the fear of a possible accident. Even today, the possibility of a "dirty" bomb detonated by terrorists or an attack on a nuclear facility is continually mentioned by the press and Homeland Security.

Some of the fears are a product of sensationalist headlines and others by coverage of anti-nuclear activists. Some of the coverage of the Chernobyl accident included a casual comparison between the Chernobyl and Three Mile Island accidents, stating that

Three Mile Island accident could have turned into a Chernobyl disaster in a matter of minutes.

Frequent arguments frequently used by anti-nuclear activists are:

- terrorist organizations could acquire spent nuclear fuel to extract enough plutonium to make an atomic bomb;
- some radioactive sub products in the spent nuclear fuel have long half lives and the need to bury them in miles-deep underground caves;
- nuclear plants are bombs under "pseudo" control; such limited control capable of slipping away in an instant, turning the installation into a much more powerful bomb (more radioactive material) which would level all within x miles radius from the plant, and;
- there is sufficient open information (on the Internet) to allow others with plenty of resources to build nuclear weapons, especially starting from spent fuel rods.

The utilities are also partly to blame for the fear surrounding nuclear plants. There have been fierce lobbying campaigns to ensure that the utilities are shielded from liability in the event of an accident and to rush through approval of the Yucca Mountain nuclear repository, arguing that spent fuel could leak or attract saboteurs.

Most of the scientific community favors nuclear energy, but the general public, rather than accepting the experts' opinion, regards their point of view as suspect because they work for the industry or the government.

No one can deny the power of atomic bombs, the destruction they produced in Hiroshima and Nagasaki and the thousands of deaths caused immediately or subsequently as a result of radiation exposure. I pray that there is never another nuclear bomb explosion.

Some of the statistics give food for thought. On average, the per capita emission in the U.S. is about 20 tons of CO_2 per year. If all the electricity were provided by coal, the per capita requirement would be about 13,300 pounds of coal per year. The ashes after combustion (solid residues, assuming 94% carbon) represent about 850 pounds per capita per year. If all electricity came from nuclear power, the total amount of nuclear waste per capita for a 70 year life span would be two pounds.

Nuclear energy today represents 8.14% of the primary energy consumed in the U.S. and generates 19.1% of the electricity. Most nuclear plants are used as base load plants. They do not emit green house gases and could quickly reduce our dependence on fossil fuels. New reactor designs can utilize nuclear energy much more efficiently, are inherently safer and produce much less nuclear waste in the form of shorter-lived radioactive material. The integrated fast reactor uses a higher uranium concentration, operates at a much higher temperature and is cooled with liquid sodium, which is risky, but rather than needing to reprocess spent fuel, continues operating through the plutonium phase, eliminating long-lived radioactive sub products.

There are about 5 million tons of economically recoverable uranium (at present prices), which at the current consumption rate could last about 80 years, with more potential reserves (exploration ceased in the 1980s as a result of the cancellation of many plants). There is plenty of uranium in the oceans, albeit extremely diluted (parts per billion), but if the price is right, there are ways of extracting it. To generate all the base load electricity needed in the US with nuclear plants (cycling and peaking plants might be needed to adjust the output, since it is better to operate nuclear plants at a constant output), about 300 new, 1GW plants need to be constructed, which would require doubling the proven reserves of uranium to guarantee the new plants could operate for 30 to 40 years.

Nuclear energy has the potential to abate green house emissions at a faster rate than renewable energy because the technology is more mature, economies of scale favor larger installations and it displaces fossil fuel base load plants.

There is no "reliable" figure to assess the cost of building a new nuclear plant. Information submitted by utilities includes large contingencies for public hearing delays, obtaining permits, and work interruptions. During the 1970s, the cost of nuclear plants was estimated at about three times that of a conventional thermoelectric plant. For the purpose of this book, it is assumed that the capital cost of a nuclear plant is ten times the cost of a state-of-the-art combined cycle thermoelectric plant, or in round numbers, about $10 million/MW.

The cost of fueling a nuclear plant, assuming U_{238} yellow cake is available at $40/lb, is about 1¢/kWh. If a nuclear plant costs ten times as much as a combined cycle plant, the overall cost of using a combined cycle plant at today's natural gas prices.

Fast breeder reactors (Gen IV reactors) utilize sodium as the cooling medium and have the potential to extract several hundred times more energy from uranium by also utilizing the nuclear energy contained in nuclear waste, which would eliminate concerns about fuel availability and reduce the size of the nuclear waste to 2% of traditional reactors and would also reduce the period during which such waste remains dangerous. Utilizing sodium as a coolant is tricky because sodium leaks are explosive.

The question is whether the public's fears are rational and whether education can change people's perceptions. Regaining trust is difficult. The possible contribution of nuclear energy in alleviating CO_2 emissions would depend on the ability of the politicians (as influencers of public opinion) and the public to change their preconceived positions.

Wind Energy

Wind energy is indirect solar energy. Wind blows from high pressure areas to low pressure areas, which are produced by different rates of absorption of thermal heat.

Mankind has been using wind energy for almost two thousand years, pushing sails, turning windmills and pumping water. During the Industrial Revolution, the utilization of wind as an energy source expanded significantly. The first installations to generate electricity were developed at the beginning of the 1900s. Wind energy was once considered the solution to the future energy crisis. However, it has proven more difficult than anticipated due to the variability of the wind. An ideal site for wind energy must have constant high winds. However, even in windy places, the wind is far from constant. Wind speed is dependent on climatological and topographical conditions. Good sites have an average wind speed greater than 10 m/s (36 km/hr).

The energy extractable from the wind varies proportionally with the cube of the wind velocity. With low wind speeds, the available energy is quite low, and a windmill is not capable of generating its rated capacity. However, when the wind starts overcoming the designed speed, mechanisms to limit the speed rotation must be brought online or the equipment shuts down. Figure No. 20 below shows power generated by equipment designed to produce 500 kW at a speed of 7 m/s.

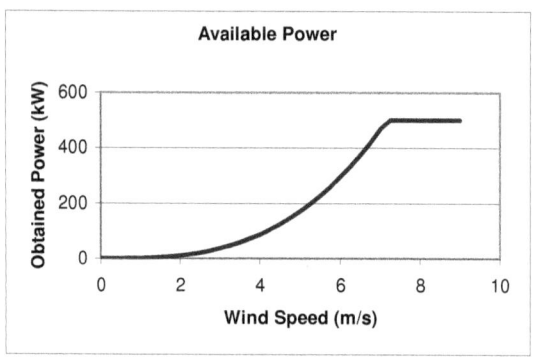

Figure 20 Delivered Power vs. Wind Speed

The graph shows that when the wind speed is 4.25 m/s, the windmill can only generate about 100 kW; at 5.25 m/s it can generate about 200 kW; at 6.6 m/s, it can generate about 400 kW. However, if the speed reaches 9-9.5 m/s, it would generate 1 MW, twice the rated capacity and likely above its safety margins.

To prevent damage to the equipment, the blades are turned to provide braking, and eventually the turbine might be locked.

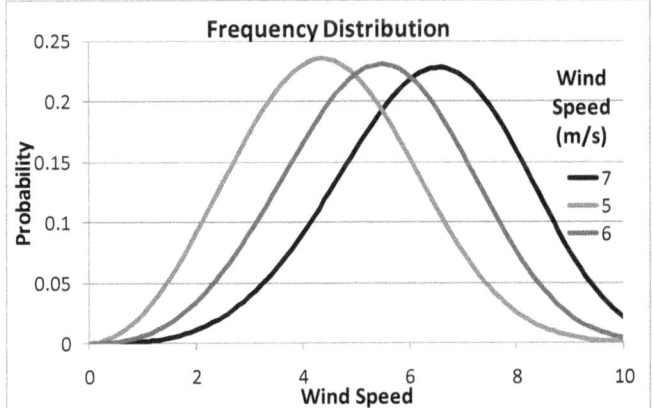

Figure 21 Wind Speed Distribution

The wind speed and direction change constantly. A plot of the frequency distribution of wind speeds over time resembles a Raleigh or Weibull distribution curve or function. Typical distribution and aggregate or cumulative distribution curves for three wind speeds are shown below in Figures No. 21 and 22.

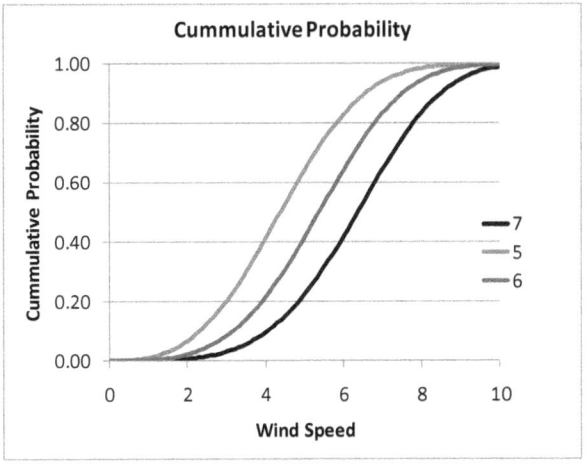

Figure 22 Cumulative Distribution

The graphs show the probability of the wind blowing at a given speed. The graph shows that for an area that has an average wind speed of 5 m/s, 18.5% of the time, the wind speed was less than 3 m/s; 45% of the time the wind speed was less than 4 m/s, 50% of the time the wind was less than 5 m/s, 65% less than 6 m/s, 76% less than 7 m/s and 88% of the time less than 8 m/s. Wind speeds between 4 and 6 m/s occurred less than 20% of the time, while winds above 7 m/s blew about 24% of the time.

The combination of these two elements - available power as a function of speed and the probability distribution of wind speeds is a lethal combination. Since the energy availability from the wind varies with the cube of the wind speed, a 6 m/s wind will provide eight times more energy than a wind at 3 m/s. The available power can be calculated by integrating the probability speed distribution times the power generated at that speed. Depending on design parameters and the site, the availability of the equipment fluctuates between 25 and 40% of the time, as shown in Figure No. 23

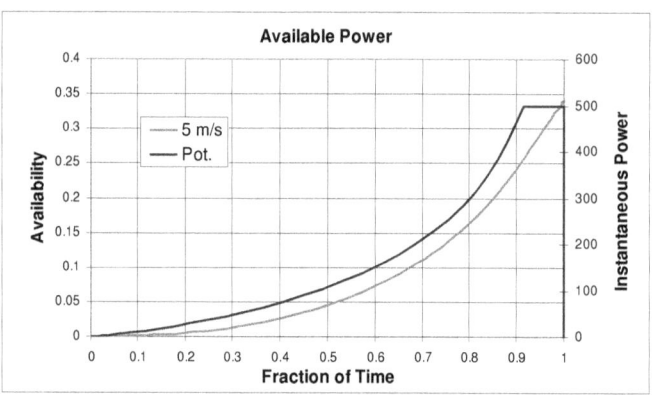

Figure 23 Available Power from Wind

The graph shows that 50% of the time the power generated by a 500 kW plant is less than 100 kW, generating less than 5% of the nominal capacity; 80% of the time, the plant generates less than 250 kW, representing only 16% of the nominal capacity. Assuming that maintenance is done during known calm days, the energy produced by this equipment would only reach 33-34% of the rated capacity. In other words, a 500 kW turbine, rather than producing about 4,380 MWh per year, would produce less than 1,450 MWh.

Large wind farms average some of the gust and short-term wind variations, and big strides have been made in load and speed management, resulting in variable but smoother generation. A few years ago, it was considered that wind energy could only complement a robust system because the system could not respond very quickly to wind variations, but spread clusters of wind farms have allowed a larger penetration of wind turbines.

Despite its limitations, today wind power is the least expensive source of electricity, resulting in an explosive growth of wind farms installed capacity worldwide, as shown in Table No. 9:

Installed Wind Capacity					
Nation	2006	2008	2010	2012	2014
China	2,599	12,210	44,733	75,564	114,763
United States	11,603	25,170	40,200	60,007	65,879
Germany	20,622	23,903	27,214	31,332	39,165
Spain	11,630	16,740	20,676	22,796	22,987
India	6,270	9,587	13,064	18,421	22,465
United Kingdom	1,953	3,288	5,203	8,445	12,440
Canada	1,460	2,369	4,008	6,200	9,694
France	1,589	3,426	5,660	7,196	9,285
Italy	2,123	3,537	5,797	8,144	8,663
Brazil	237	339	932	2,508	5,939
Rest of the World	14,055	20,619	30,150	41,869	58,279
Total World	74,141	121,188	197,637	282,482	369,559

Table 9 World Wind Energy Installed Capacity (MW)

The growth rate has been spectacular. In 1981 there were 25 MW installed; in 1991, 2170 MW; in 2000, 17,300 MW; in 2005, 59,000 MW; in 2010, 196,600 MW and at the end of 2014 there were 370,000 MW installed. The growth rate in China in the last eight years has been 551%, in Brazil 313%, in the US, Canada, UK and France above 70%, the rest of the world 51.8% and the whole world 62.2%. Spain and Germany are more mature markets and their growth rates have been lower, but the penetration of wind in the total generation hovers around 15%. The ten largest wind energy-producing countries had an average availability of 20%, but in 2008, it peaked at 24.5%.

Wind represented 16% of the electricity generated in Spain in 2011, 16% in Ireland in 2012, and 30% in Denmark in 2014, with about 90 hours that year that wind produced all or even a bit more than all of the electricity needed. In the US, wind generation represented less than 4.0%.

Despite its privileged position as the least expensive form of electricity, wind energy is the least reliable source. Even in Altamont, California - with "predictable" wind patterns as air moves from the ocean inland during the day. Daily, hourly and even shorter interval variations require thermal equipment be ready on standby to cover quick changes. Wind energy is a great resource for non-sensitive time performance (e.g. pumping water into a tank or reservoir).

Wind energy would be a good complement for a robust system. Separation of wind farms or a cluster of wind farms together with geographical diversification could allow a larger penetration in the US, possibly reaching 10-15% of all the electricity generated in the country. Since electricity consumption in the country is 4.1 PW/y, 15% means about 615 million MWh/y generated by wind turbines. To reach that level, and assuming an availability of 25%, would require installing about 280,000 MW, or about five times the current installed wind capacity.

Environmentalists have opposed wind energy because it is noisy, kills some birds and spoils the view. Earlier wind turbines were noisier and less efficient. Improved efficiencies have resulted in a lower noise level. Typical noise levels in rural settings range from 20-40 dBA), similar to the noise level of a wind farm 350 meters away (35-45 dBA), but substantially less than the typical noise in a normal busy office (about 60 dBA). Bird fatalities have also decreased with larger diameter, slower turbines, wider spacing and the recognition of the need to preserve some corridors for local fauna. The modification of the landscape is subjective. If there were no other alternatives and the choice were between having electricity or preserving the landscape, most people would volunteer to go and install the wind turbines themselves. With other alternatives – e.g. a nuclear plant or a thermoelectric plant - (preferably out of the state), preserving the landscape might be possible, but it would be selfish. Developers choose the sites carefully, taking into consideration wind densities, but they probably would settle for a site with marginally less wind if they would thereby avoid a lengthy environmental legal battle.

Photovoltaic Panels

This is the most modern form of capturing energy from the Sun. It is very simple, has no moving parts requiring maintenance and it even looks pretty and sophisticated. It is totally scalable and can be constructed to give output from microwatts to power watches to megawatts to power large applications. PV panels (a panel is an array of cells) have evolved from a laboratory curiosity into a multibillion dollar market.

The principle is very simple. A photon (packet of energy) in the light can dislodge an electron when it hits an atom. Different atoms hold onto electrons with different strengths depending on the number of electrons in the atom's outermost shell. Some elements on the left side of the periodic table will give up electrons easily (called electropositive), while elements on the right hand side are willing to grab whatever electron come their way (called electronegative). When an electropositive element combines with an electronegative one, they form an ionic bond, and the two elements are held together by the attracting charges. Elements in the middle of the periodic table tend to arrange themselves in crystalline structures in such a way as to share electrons among themselves, forming weaker covalent bonds.

Since the covalent bond is weaker, lower energy photons (like those in the visible spectrum) can dislodge electrons. The strength of the covalent bond determines whether the material is an insulator (strong covalent bond), a conductor (weak bond) or semi-conductors (in between). Silicon is the most used semi-conductor. Silicon has four electrons in its outer shell. Pure silicon atoms arrange themselves to form a stable crystal structure, sharing two electrons with each of four neighbors. If phosphorous, which has five electrons in its outer shell, is introduced as an impurity in the silicon, the "doped" material seems to have extra electrons (even when it is electrically neutral). This doped material is called an n-type semiconductor. If boron, which has three electrons in its outer shell, was introduced as an impurity in silicon, it appears that the doped material is lacking one electron (even when it is electrically neutral). Such doped material is called a p-type semiconductor.

When an n-type semiconductor is placed in contact with a p-type semiconductor, there is an instantaneous transfer of electrons from the "excess" n-type material to the "holes" in the p-type material. As a result, the n-type material becomes positively charged and the p-type material receiving the electrons becomes negatively charged. The negatively charged p-type material restricts any further flow of electrons from the n-type material, while at the same time the positively charged n-type material attracts electrons. This restriction makes the transfer of electrons go only one way, from the n-type to the p-type.

When a photon of light strikes an electron, the energy of the electron gets increased by the energy of the photon. If the amount of energy is above the energy required to move the electron to a higher orbit, the electron can then move freely. If the energy is below the level required to reach the higher orbit, the energy only increases the kinetic energy, which manifests as increased temperature. The amount of energy needed to propel one electron to a higher orbit is called the band gap. If the photon has more energy than the band gap, the freed electron will take the excess as kinetic energy.

In a photovoltaic device, as free electrons are generated on the n-layer, the electrons can either go through the external circuit or move towards the p-layer, where, due to the negative charge of the layer, they are repelled. If the n-layer is very thin, the possibility of recombining within the n-layer is small and most electrons will go through the external circuit. If the external circuit is open, the electrons eventually recombine with the holes in the n-layer, resulting in a temperature increase of the device.

Different elements or materials have different band gaps. The band gap is normally expressed in electron volts (eV). The energy of a photon is related to its frequency. Photons with shorter wavelengths (higher frequency) near the blue end of the visible spectrum have higher energy than those photons with longer wavelengths near the red end of the visible spectrum. Solar radiation with wavelengths less than or equal to the corresponding band gap will be absorbed by the material. Solar radiation with wavelengths larger than the band gap will pass through the semiconductor mostly undisturbed. Silicon has a band gap of 1.11 eV, corresponding to a wave length of 1120 nm, which allows it to capture a good portion of the visible spectrum.

Electrical Characteristics of PV materials

The performance of a typical photovoltaic material ("cell") can be assessed simply with an ammeter, a voltmeter and a variable electric resistance. When the resistance is infinite (i.e. the cell is not connected), the current is at its minimum and the voltage at its maximum, known as the open circuit voltage (V_{OC}). At the other extreme, when the resistance is zero, the cell is in effect short circuited and the current its maximum, known as the short circuit current (I_{SC}). The performance of a cell as a result of varying the resistance is shown in Figure No. 24 below:

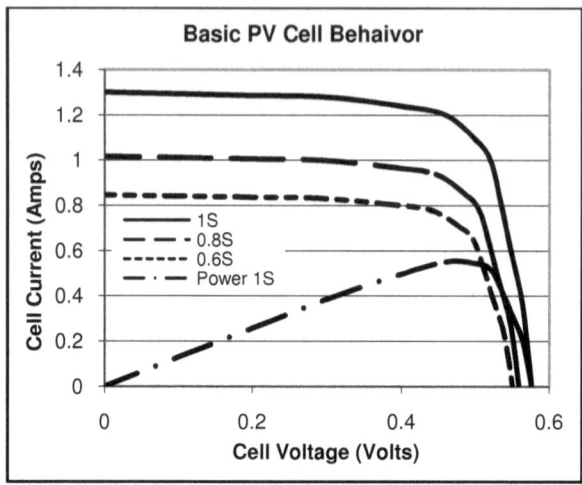

Figure 24 PV Cell Behavior

Exposing the cell to different light intensities produces the similar curves shown in Figure No. 24, with the short circuit current I_{SC} decreasing proportionally to the intensity of the radiation, while the open circuit voltage falls less sharply. The power output of the cell (the product of the voltage times the current) is zero at I_{SC} and again at V_{OC}. Between these two points, the power rises and then falls, and there is a point where maximum power is delivered. The power output curve for only the largest solar intensity is shown as the broken line in Figure No. 25. Similar power curves can be drawn for other solar irradiation intensities.

As mentioned in Chapter 6, the intensity of solar irradiation changes continuously as the sun moves across the sky. Thus, a solar cell or a PV panel will produce little electricity during the early morning and late afternoon hours, reaching a peak around noon. Furthermore, the output of a PV panel is affected by temperature. The effect is rather constant and is normally expressed as a percentage change per °C and, depending on the material and manufacturer, can normally range from -0.3%/°C to -0.5%/°C. The negative value means that as temperature increases, the output decreases.

To be able to compare two PV panels, there is an international agreement to report the power of a cell in peak watts (w_p), which measures the output at "standard testing conditions" of 1,000 W/m^2 hr, air mass of 1.5 and 25°C. The standard is equivalent to reporting gas mileage for cars at standard testing conditions of 80 mph on a windless day at 25°C, but going downhill. Few minutes a year are close to these standard conditions. Fortunately, two similarly rated panels, placed in the same location, should be able to produce a similar amount of electricity during the year. Table No. 10 compares two almost equal sized panels, a thin film amorphous panel and a single crystal panel, which have slightly different efficiencies and prices.

	Price	Area	Wp	Efficiency	kw/y	Cost/kWp
	$/panel	square m²	w		per m²	$
Panel A (multicrystal)	228.80	1.642	260	15.8%	239.30	880
Panel B (single crystal)	308.00	1.696	280	16.7%	249.50	1,100

Table 10 PV Panel Output

It is worth nothing that the price of PV panels has dropped considerably in the last decade, and now it is possible to purchase panels for about $1/watt. The prices today are about 1% of the price 40 years ago and 10% of the prices a decade ago.

The actual power delivered by a PV panel or array of panels depends both on the characteristics of the site and of the panel. The most important ones are:

- solar irradiation available at the site;
- orientation and inclination of the PV panel;
- efficiency of the PV panel, and;
- sensitivity to temperature changes.

To calculate the possible output of a PV panel at a given location requires detailed meteorological data and an estimation of the cell temperature under those meteorological conditions. A quick approximation is to use the efficiency of the panel as a proxy and apply a correction factor for the degree day departure from 25°C. The output of one peak kilowatt PV array in San Diego is about

1,500 kWh/y, while the output of one peak kilowatt PV array in New York is about 1,200 kWh/y.

PV panel demand has exploded in the last few years, mainly in Germany and Japan, propelled by incentives to entice users to install residential systems. At the end of 2014, a total capacity of more than 178 Gigawatts (GW) had been installed worldwide, as shown in Table 11.

Year-end	2010	2011	2012	2013	2014e
Cumulative	40,336	70,469	100,504	138,856	178,391
Annual	17,151	30,133	30,011	38,352	40,134
Growth p.a.	74%	75%	43%	38%	28%

Table 11 PV Installed Capacity Worldwide (MW)

A breakdown of the installed capacity and new installations in the year 2014 is shown in Table No. 12.

	Total Capacity (MW)			**New Installations (MW)**	
	Country	**Capacity**		**Country**	**Addition**
1.	Germany	38,200	1.	China	10,560
2.	China	28,199	2.	Japan	9,700
3.	Japan	23,300	3.	United States	6,201
4.	Italy	18,460	4.	UK	2,273
5.	United States	18,280	5.	Germany	1,900
6.	France	5,660	6.	France	927
7.	Spain	5,358	7.	Australia	910
8.	UK	5,104	8.	South Korea	909
9.	Australia	4,136	9.	South Africa	800
10.	Belgium	3,074	10.	India	616

Table 12 Breakdown of Installed PV panels

Given the solar irradiation available in Germany and Japan, PV panels should be capable of producing about 1,000 kWh/y for each kW_p installed, which represents about 6.0% and 2.1% respectively of the total demand for electricity in each country.

PV Limitations

There are three limitations with PV panels. The first relates to the energy required to construct them, which has decreased significantly with the introduction of amorphous silicon and is now around one year of generation. The second limitation relates to cost, which, with mass production and thinner wafers, has decreased from $10-15 per peak watt to about $1/$W_p$ today (about $2.5/$W_p$ installed), with high expectations that the cost might continue decreasing in the foreseeable future. The third limitation is critical. PV panels do not operate during the night. They are great for remote locations or places with restricted access, where providing a battery for storage is economically justified. For residential, commercial or industrial applications, their economic benefit, both for the individual and for society, is limited.

As was shown before, consumption of electricity is highly variable both at the residential and grid level. A customer wanting to become independent of the power grid would have to include storage to allow night operation and would have to substantially modify his consumption habits. An easy solution sold to the public is to direct them towards becoming self-sufficient on a cumulative basis, using net metering which requires a new meter that can move back and forward. If a customer can generate 24 kWh during the day and deliver that to the grid, he/she can consume on average 1 kWh during the day and would not have to pay for electricity because the meter was reversed. The user would be using the grid like a big battery and storing excess energy there, drawing it when needed.

The output of a PV panel depends on the strength of the solar radiation and thus it peaks around noon time, which unfortunately does not coincide with the peak demand for electricity, which means that PV panels displace a cheap source of electricity, namely the base load plants. At low levels of penetration, there is no problem, but if PV panels were to generate 10% of the base load, base load plants would have to reduce their generation capacity during the day but during late afternoon, peaking plants will still be needed to cover demand.

The PV panel's great advantage is its solid state composition (no moving parts) and its scalability. Its main disadvantage is the need for storage. The uninstalled cost of one MW of peak power is about $1 million. A state-of-the-art thermoelectric plant also costs about $1 million/MW, but it can generate in one year more than five times the electricity of PV panels for the same price. Wind energy is today about 10 times cheaper than PV electricity and while neither alternative can operate continuously, wind electricity is already price competitive.

Ocean Energy, Geothermal and Hydro electricity

The three distinct forms of renewable energy are lumped together because the availability of these resources is dependent on their geographical location.

Hydroelectricity is a well known source of electricity. It has been exploited for more than 100 years and will likely continue to be a reliable and inexpensive source. Although the most attractive sites have already been tapped, it might be possible to double the output in the U.S. by exploiting second-tier resources. Since hydroelectricity represents only 2.4% of the primary energy demand in the U.S., doubling the output might help but will not provide a solution to the energy problem.

Geothermal energy is the only renewable energy that is not dependent on the Sun. There are many places where the heat of the Earth's core is sufficiently concentrated in proximity to the surface, producing hot water springs and/or geysers that can be tapped to produce electricity. The U.S. has an installed capacity of about 2,500 MW of geothermal energy, tapping the most obvious high temperature areas. It is not a resource to be "mined"; it has to be carefully exploited, otherwise the resource is cooled and losses its potential. New ground source heat pumps, taking advantage of the temperature increase with depth, are starting to contribute to heating/cooling for residential sized units. Generation of electricity might be possible, but the temperature difference is probably too small.

Tidal energy comes from the continuous rise and fall of the oceans caused by the gravitational pull of the Moon. The overall effect of the Sun/Moon interaction is a tidal range in the middle of the oceans of about 0.5m. The tidal range experienced at coastal sites can be substantially modified by local topographic conditions and can reach, in some locations, up to 10-15 m. A 240 MW barrage/dam was built in France in the 1960s and has operated successfully for many years converting tidal energy to electricity. There are a few locations in the U.S. that have tidal ranges in excess of 7 m which could be exploited. There are many projects on the drafting board in the world, but the high cost and risks are preventing their development.

Wave energy attempts to capture the energy available on waves. There are many designs and experimental stations that have been undergoing testing, some for many years, some aiming to capture the energy as waves hit the shore while others aim to capture the energy produced by the up/down swells in deep water. Although the potential is high, costs of generation energy with waves are still too high and the resource does not allow for mass production since chambers/depths are usually tailored to specific wave patterns at the desired location.

The ocean thermal energy conversion ("OTEC") attempts to exploit the temperature difference between hot, surface tropical waters and cold, deep water to generate electricity with a typical Rankine engine. Since the temperature difference is quite small (about 20°C), OTEC requires moving rivers of water in huge equipment, and thus, investing in this potential source of energy was not attractive in the past, but could be competitive today. It is a tropical resource requiring water depths of about 1,000 m, which is not readily available on the continental U.S., but could be exploited in Hawaii. There was a small land-based experimental station in Kea Hole. One of its main advantages is that it could work continuously (except during storms), and preliminary calculations suggest an investment of about $3-4 million per MW would produce electricity at competitive prices today. Since there is no consumption of electricity in mid ocean, deep water near the coast is a requirement.

Bioenergy

Bioenergy refers to energy obtained using materials that were produced from biological sources, whether as energy crops or as organic waste. Such materials can be burned directly to produce heat or power or can be converted into biofuels, usually in more developed countries. Biofuels refers to the products resulting from the transformation of some energy crops into a liquid fuel that can be used for transportation.

There is no precise information of the contribution of bioenergy to the total energy consumption. Many sources, for example burning firewood or dung, account for a large part of the energy consumption in many poor developing countries and are usually not reported. Some others are estimated, like the use of residues in the pulp and paper industries, in sawmills, or even bagasse burning in sugar mills. A United Nations estimate considers that traditional use of biomass for energy represents about 10.6% of the primary energy consumption in 2000, which works out to about 1,100 mtoe (million tons of oil equivalent). New biomass (i.e. energy crops) represents today about 200 mtoe.

Bioenergy, as a source of energy, can be classified as energy crops and utilization of wastes. The utilization of wastes includes the use of residues from energy crops and food crops, the treatment of animal manure, sewage sludge, municipal solid waste and landfill gases. Given the compelling evidence that generating electricity from wastes is highly economical and beneficial, it would be ill advised to recommend that more waste be created to increase the electricity generation, and therefore will focus the discussion of bioenergy only on energy crops.

The yield, the weight of biomass produced per unit area per year, is critical for energy crops. It depends on many factors: location (weather, climate, water availability and solar irradiation), the nature of the soil, the amount of fertilizer required and the type of plant. Depending on the plant, the yield can range from 1 ton/ha to 30 tons/ha. In terms of energy, the range is between 15 GJ/ha to 300 GJ/ha. For comparison, oil has about 42 GJ/ton and a hectare receives yearly about 20,000 GJ/ha from the Sun. Therefore, in the best case scenario, only about 1.5% of the solar energy is being extracted as an energy crop. The energy content per weight of dry biomass is about 35% of the energy content of oil and about 25% on a volume basis.

Ethanol

Ethanol can be mass-produced from the sugars and starches of a wide variety of crops, such as sugar cane, miscanthus (a grassy plant), sugar beet, sorghum, switch grass, barley, hemp, potatoes, cassava, sunflower, fruit, corn, wheat and cotton, as well as many types of cellulose waste and harvesting byproducts. The basic steps for large scale production of ethanol are the fermentation of sugars and distillation, although cellulosic crops require prior to fermentation the saccharification or hydrolysis of carbohydrates such as cellulose and starch into sugars. The crops are grown, collected, dried and processed, all of which require manpower, equipment and energy. The energy ratio represents the relationship between the total energy needed for the processing (from fertilizer to transportation of products) to the energy released by burning the fuel.

Table No. 13 shows worldwide production of ethanol (ethyl alcohol) for fuel. It has reached some 24.6 billion gallons, produced mainly from corn in the US and from sugar cane in Brazil. Brazil and the US together contribute almost 80% of the ethanol produced as fuel[10].

Production of corn ethanol increased drastically in the US thanks to subsidies, just as Brazil offered in the early 1980s. It is important to notice that the yield of ethanol per hectare and the energy ratio is substantially higher for sugar cane than for corn. The penetration of ethanol in the gasoline market represents about 9.8% in the U.S. and more than 50% in Brazil.

---

[10] www.ethanolrfa.org/pages/annual-industry-outlook (August 2015)

Country	2007	2009	2011	2013	2014
World Fuel Ethanol Production by Country or Region (Million Gallons)					
USA	6,521	10,938	13,948	13,300	14,300
Brazil	5,019	6,578	5,573	6,267	6,190
Europe	570	1,040	1,168	1,371	1,445
China	486	542	555	696	635
Canada	211	291	462	523	510
Rest of World	315	914	698	1,272	1,490
WORLD	**13,123**	**20,303**	**22,404**	**23,429**	**24,570**

Table 13 Worldwide Ethanol Production

A comparison of the yield of ethanol that can be produced by sugar cane and corn is presented in Table No.14:

Characteristic	Brazil	U.S.
Feedstock	Sugar cane	Corn
Total production in 2014 (million gallons)	6,190	14,300
Total arable land (million ha)	355	270
Total area used for ethanol crop in 2013 (million ha)	4.8	22.4
Yield per hectare (liters of ethanol/ha)	6,800-8,000	3,800-4,000
Ratio of energy (energy in ethanol divided by energy needed to produce ethanol)	8.3 to 10.2	1.3 to 1.6

Table 14 Comparison of Ethanol Production (2014)

Ethanol can be used directly in vehicles prepared to utilize it or can be blended with gasoline, up to about 10% without having to modify or even tune a normal car. With advanced electronics and corrosion resistant materials, some new cars are able to use a higher percentage of ethanol. The mileage attainable with ethanol is less than that of gasoline (ethanol has lower density and heat content than gasoline) and is sold in Brazil at a discount over gasoline prices, so that the cost per mile driven is similar.

The use of ethanol as fuel has faced strong opposition from many people, who claim that:

- biofuels caused world food prices to increase by 75% because producing ethanol consumed one third of America's corn;

- filling up one fuel tank one time with 100% ethanol uses enough corn to feed one person for a year;
- incorporating externalities into the production of corn ethanol results in a negative energy input;
- cellulosic ethanol from switch grass and saw grass, removes nutrients and might result in erosion;
- the land requirements for ethanol to substitute gasoline would exceed the available cultivated land.

Some of these opinions are expressed with great passion, but are debatable. As ethanol yields improve using biotechnology to process the cobs and fodder, ethanol production may become more attractive. If oil prices were to continue at around $150/barrel, it would be viable to use other feed stocks, such as cellulose. Some fast growing species, like switch grass, can be grown on land not suitable for other cash crops, with high yields of ethanol per unit area, and fertilizer could replenish some of the lost soil nutrients.

Rather than participating in the discussion about the merits of energy crops, a comparison to other ways to capture solar energy provides a sobering reality. In the US the average yield for corn is about 170 bushels per acre (about 11.6 tons/ha). Each bushel can be converted into 2.5 gallons of ethanol, resulting in about 425 gallons per acre per year (about 4,000 l/ha – 3.2 tons/ha). The energy captured by corn ethanol can be calculated from the above numbers, as follows:

$$\text{Energy (w)} = \frac{170 \text{ bu/ac y} * 2.5 \text{ gal/bu} * 76,000 \text{ BTU/gal} * 1,054 \text{ J/BTU}}{4,000 \text{ m}^2/\text{ac} * 31.5 \times 10^6 \text{ s/y}} = 0.27 \text{ W/m}^2$$

For comparison, a multi-crystal PV panel with 10% overall efficiency in San Francisco with average solar irradiation of 17.6 MJ/m^2 d captures:

$$\text{Energy (w)} = \frac{\text{Solar irradiation } 17.6 MJ/m^2 \text{ d} * \text{efficiency } 10\%[11]}{86,400 \text{ s/d}} = 20.4 \text{ W/m}^2$$

---

[11] After inverters, transformers, metering, etc.

The conclusion is that a PV panel can produce 75 times more energy than corn per square meter. Granted, ethanol is a liquid fuel that can be used directly as fuel and a PV panel produces electricity. However, if we assume that the overall efficiency of converting electricity into hydrogen is 50%, producing hydrogen will still be 37 times more efficient than producing ethanol.

Solar Thermal

Solar thermal refers to the use of solar rays to heat air, water or other substances. It includes many applications, ranging from passive heating of houses to high concentrating collectors utilizing molten salts as the heat carrying medium.

Depending on the final temperature reached by the system, they are classified as:

- Passive heating, which seeks a more efficient capture of the Sun's heat to maintain the temperature of the room/house at comfortable level, thus reducing the consumption of other energy sources.
- Water heating, which is used to provide hot water for a residential unit, normally utilizing a flat collector which can increase the water temperature to slightly below boiling. In regions with no danger of freezing temperatures, a person can install a simple thermo siphon system combining storage and heating elements, while in regions subject to freezing, a pumped system (with anti-freeze and/or a mechanism to empty the collector) is required.
- High temperature, which requires the use of concentrating collectors and Sun tracking mechanisms to reach higher temperatures and improved efficiency.

Passive heating is easy to implement in new constructions and should become part of new building codes. It can save owners a substantial amount of energy and does not result in more expensive houses. It is more difficult to retrofit existing houses. Passive heating does not produce energy, but it does save it.

Water heating was very popular prior to World War II. In Miami, 80% of houses built between 1935 and 1941 had solar heaters. As with passive heating, solar water heating does not produce energy, but can reduce the need for fuel and/or electricity. Their simplicity and the availability of backup fuel or electric heaters should make it mandatory in all houses.

High temperature systems

These systems are capable of producing electricity today at near competitive prices with some incentives.

To reach high temperature, the incident solar irradiation has to be concentrated. The most common approaches are with trough parabolic collectors, parabolic dish concentrators and power towers using Sun-tracking heliostat mirrors that reflect the Sun's rays onto a boiler situated on top of a tower. The high temperature is needed to improve the overall efficiency of the system.

Between 1984 and 1990, Luz International developed nine large power stations (called SEGS I through IX) at the Kramer Junction in the Mojave Desert, generating between 13 and 80 MW, with a combined power of 354 MW. Large trough parabolic collectors tracking the Sun heat synthetic oil to 390°C, which in turn heats water to move a steam turbine. Given the commitment to deliver electricity, to guarantee throughput and improve efficiencies, the systems rely on gas as a backup. Overall, around 60% of the generation is from solar energy. Although Luz experienced financial difficulties in 1992, when electricity prices, tied to the price of natural gas decreased, the systems (with new owners) have continued operating. Solar One, rebuilt as Solar Two at Barstow, California, operated a 10 MW tower plant utilizing molten salts at over 500°C, and a new tower plant being built in Sevilla, Spain, will operate with air at almost 700°C. There have been several designs using parabolic dish concentrators, utilizing a Stirling engine at the focal point of the dish, but none has reached commercial operation.

The concept of concentrating the solar irradiation to achieve higher temperatures is sound, and in the last few years, there has been a surge of new projects. Ausra, utilizing Fresnell concentrators built a 177 MW installation; Bright Source Energies built a tower illuminated by thousands of heliostats for a 400 MW plant on the California-Nevada border; and Solel, an Israeli company, signed a power purchase agreement with PGE for a 553 MW plant in California. These projects are becoming economical at what is known as the feed-in price, which is the cost that utilities are paying for electricity generated by a new combined cycle plant. These projects have tripled the existing capacity in the Mojave Desert.

Although those plants do not provide for much thermal storage of energy, by seeking operation at higher temperatures to improve efficiencies, they postpone the generation of electricity until the set temperature is reached, and thus the systems are actually shifting the generation towards the afternoon, making a larger contribution to peak hour demand.

The strategy makes sense – to compete with fossil fuel power generation, economies of scale and high temperatures are needed. To reach high temperatures, the concentration ratio needs to be high, which requires mirrors with smaller openings and thus requires tracking the Sun, resulting in a tradeoff between high efficiency and complexity.

The plants in the Mojave Desert have been an operational success. They have been able to deliver the contracted power for many years. The use of natural gas to substitute for solar energy on cloudy days has worked better than expected. Some of the operating characteristics provide interesting conclusions, as shown in Table No. 15 below:

Characteristics of SEGs Systems							
Plant		I	II	IV	V	VII	IX
Nominal Capacity	(MW)	13.80	30.00	30.00	30.00	30.00	80.00
Production/year	(MWhr/y)	30,100	80,500	91,311	99,182	92,646	256,125
Gas Consumed	(b m^3/y)	4.76	9.46	9.63	10.53	8.10	25.20
Gas Generated	(MWhr/y)	18,519	36,804	37,465	40,967	31,513	98,040
Solar Generated	(MWhr/y)	11,581	43,696	53,846	58,215	61,133	158,085
Solar Contribution	%	38.5%	54.3%	59.0%	58.7%	66.0%	61.7%
Efficiency Improvement		1.63	2.19	2.44	2.42	2.94	2.61
Hours Operation		6.23	7.67	8.70	9.45	8.82	9.15
Hours Operation Solar		2.40	4.16	5.13	5.54	5.82	5.65
Equivalent Capacity	8000 hrs	3.76	10.06	11.41	12.40	11.58	32.02
Solar/Thermal		0.63	1.19	1.44	1.42	1.94	1.61

Table 15 Operating Characteristics of SEGS Plants[12]

The table shows that the solar contribution to generation has increased from 38.5% to 66% on SEGS VII. The subsequent drop on SEGS VIII and IX to 61.7% was in response to economic incentives to guarantee a given capacity. The average number of hours of plant operation (calculated by dividing the yearly production by 350 days per year and the nominal capacity) has reached almost 9.5 hours per day on average. The equivalent capacity would be the rated capacity of the unit, assuming it was in operation 8,000 hours a year, like a conventional plant.

Using solar energy to complement the fuel of a thermoelectric plant would improve the efficiency of the system, decreasing the heat rate. The efficiency of the thermal section is reported to be 37.6%. If the SEGS VII system were to operate 24 hours, of which 5.82 hours would have an efficiency multiplying factor of 2.94 and the other 18.18 hours the normal efficiency, the average multiplying factor would be 1.47, which is equivalent to an overall efficiency of 55%, about the efficiency of a combined cycle plants nowadays.

These figures are impressive. They demonstrate that solar-based energy could be competitive with fossil fuel electricity when the price of natural gas is $2/MBTU. It also shows some limitations. Despite the financial strength of Luz International and the quality of the technical team, they choose not to demonstrate that it is

---

[12] Based on data provided by Goswami – Principles of Solar Engineering – Taylor Francis 2000

possible to generate electricity continuously because it would have been uneconomical.

A more modern plant is Andasol, in Spain, a new generation solar thermal parabolic trough plant, which stores energy in molten salts and provides an availability of 41%. It requires almost 10,000 m^2 of collectors per MW of nominal generation and some 28,500 tons of nitrate salts to store the thermal energy. Andasol is a 50 MW plant which cost $380 million (about $7.6 million per nominal MW).

Mid-temperature systems

A strategy being proposed by Solera is to operate a much simpler system at mid-temperatures, sacrificing efficiency for simplicity and including a way to store energy in the form of pressurized hot water, to allow continuous operation, or even better, as a peaking plant, thus aiming to compete with more polluting and less efficient peaking plants, rather than with efficient base load plants. Although economies of scale will always favor large installations, solar energy still requires the same area of collectors irrespective of the scale, permitting the installation of small distributed energy systems applicable to small communities or green buildings. The solar energy produced by a distributed system would have to compete with tariffs at the distribution level and not at the generation level.

Neither the high temperature nor mid-temperature systems could guarantee generation 24/7, 365 days a year. Both of them would have to provide (if the grid is not available) backup systems, either in the form of gas turbines for cloudy days or by heating with gas the medium that should have been heated by the Sun. If in a given location, gas was used 60 days per year, there would be 300 days when no fossil fuels are needed.

The raw estimated cost of a low efficiency system capable of operating continuously today would exceed $15 million/MW, because the cost is based on the investment divided by the rated capacity. The same system, capable of generating the same amount of electricity per day or per year, connected to a larger turbine and operating as a peaking plant, would cost less per rated capacity ($4 million/MW). Go figure!

Hydrogen

Hydrogen has been heralded as the solution to all our energy problems. It is a clean source of energy, produces only water as exhaust and is abundant. While these statements are correct, the only caveat is that there is no free hydrogen (no deposits of hydrogen exist). Although abundant, it must be separated to become a fuel, and just like electricity, it is not a source of energy but a carrier of energy.

Thermodynamically, it requires the same amount of energy to separate hydrogen from water as the energy that would be produced as a fuel, reacting with oxygen to produce water. In practice, because of efficiency losses, it would take more energy to separate than the energy it would deliver. As for clean energy, it depends on how the hydrogen is produced – if we use natural gas to do it, we also produce CO_2. Using non-polluting energy to produce hydrogen would eliminate the production of CO_2. If we envision using wind energy or photovoltaic energy to generate electricity which in turn produces hydrogen, we could have a clean CO_2 free source. The following sections discuss some of the current issues.

Transportation and Storage

There is a large hydrogen production worldwide. About 50% of the hydrogen produced is used to make ammonia for fertilizers. Early methods of production consisted of steam reforming and electrolysis. Steam reforming originally used steam with coal (partially burning coal with a controlled amount of oxygen) and then having the steam react with carbon monoxide. Electrolysis of water uses electric energy to split its constituent hydrogen and oxygen atoms. Today, the preferred method to produce hydrogen is using natural gas and steam. The reaction is reversible and endothermic. Other, newer approaches like bio-generation, photochemical or thermo-chemical reactions or photo-electrolysis are still laboratory curiosities and currently quite expensive.

Hydrogen is the lightest gas. Although it can be liquefied, the liquefaction temperatures is quite low (-253°C). Hydrogen is also highly explosive. Mixtures of hydrogen and air, ranging from 4% to 79% of hydrogen are explosive. Fortunately, hydrogen disperses quickly. Transportation and storage are problematic. Hydrogen can be pumped through natural gas pipelines (some modifications would be needed in the seals and compression equipment) to deliver it to consumption points. It is easier and cheaper to transport electricity, via cable, than hydrogen by pipelines and have the hydrogen generated on site by hydrolysis.

Storage is another problem. Given hydrogen's low density, it would require either large tanks or high pressures. Let's assume a typical gas station selling 15,000 gallons of gasoline a day decides to convert to selling hydrogen. Storing it as liquid hydrogen (LH2) would require a sphere with a radius of 3.75m, storing it as gas at very high pressure (8,000 psi) would require a sphere of 4.3 m radius or at high pressure (2,000 psi) a sphere of 6.8 m radius or at low pressure (500 psi)[13]a sphere of 10.8 m radius. The oil industry uses high pressure spheres extensively, but not in these dimensions or high pressures.

Hydrogen as a liquid fuel

Hydrogen can be used directly in an internal combustion engine (small changes to the carburetors, spark plugs, etc.) with greater efficiency than gasoline. However, thermodynamically, it would be much more efficient to use the hydrogen in a fuel cell to generate electricity to move an electric motor. Again, the main problems are associated with storing hydrogen in the car and in the best way to deliver the hydrogen to the consumer. Hydrogen can be stored in liquid form at -253°C, requiring substantial insulation that would increase the size of the storage container. Losses

---

[13] The energy content of hydrogen, on a weight basis, is similar to gasoline, but not on a volume basis. With low pressure, hydrogen expands and occupies more volume.

due to vaporization (as high as 2% per day) must be vented to prevent an explosion. Alternatively, it can be stored under high pressure, chemically in metal hydrides, or generated in situ with iron-steam oxidation. Table No. 16 illustrates some characteristics of the options available.

Hydrogen Storage Density and Tank Cost			
Form	Density MJ/l	Density MJ/kg	Cost of tank ($)	Time to refill the tank (min)
Liquid Hydrogen Dewar	5.0	15.0	1000-2000	>5
Compressed at 8000 psig	3.4	7.0	4000	2-3
Fe-Ti Hydride	2-4	1-2	3500-5500	20-30
Fe Steam Oxidation	5.8	5.0	500	2-3
Gasoline(for comparison)	30.1	24.1	20	2-3

Table 16 Hydrogen Storage in a Car[14]

The iron-steam oxidation does not store hydrogen, but allows to be generated on demand. Compared to gasoline, hydrogen is at a tremendous disadvantage. Probably the most economical method of having hydrogen is the iron-steam oxidation but it still has a lot of problems, including weight, the multiple steps of the oxidation and the separation of spent fuel. The method is really a re-generation method. Finely pulverized iron is oxidized with steam to form ferric oxide, liberating hydrogen that can be fed to a fuel cell. The oxidized ferric oxide has to be recycled to regeneration/reducing unit where the oxide is reduced again to iron, with hydrogen. Additional research is needed to simplify the substitution of spent ferric oxide for reduced iron in a practical and economical way. Future gas stations might either replace the oxidized iron with a vacuum system or provide a new standardized container with reduced iron. Customers are accustomed to having gas stations available almost everywhere and to purchasing as much as they want without having to worry if there is some gasoline left in the tank. Substituting a 50 - 200 kg tank or substituting the spent iron might be doable, but crediting the consumer for partially oxidized iron might be more complicated. In situ reduc-

---

[14] Renewable Energy – Sources for Fuels and Electricity – Johansson – Island Press - 1993

tion of spent iron oxide is probably not practical. The market-ing/distribution challenges are big. Infrastructure needs to be de-veloped to produce, distribute and dispense hydrogen.

Cost of Producing Hydrogen

In addition to solving some of the problems mentioned above, another critical element is the cost of producing hydrogen. One way of generating hydrogen without also generating CO_2 is by electrolysis with renewable electricity.

The heat of combustion of hydrogen is 57.8 kcal/mole or about 120 MJ/kg. Assuming an overall efficiency of 65% (including compressing the gas to a reasonably high pressure), the amount of energy needed to separate hydrogen from water is about 200 MJ/kg, which is equivalent to about 55.5 kWh/kg. The energy content of one gallon of gasoline is roughly equivalent (95%) to one kg of hydrogen. The cost of generating hydrogen depends on the price of electricity. Table No 17 below shows the cost of producing hydrogen, depending on the price of electricity and is expressed as the equivalent price of one gallon of gasoline and as a cost per mile driven taking into account a higher (25%) ex-pected efficiency[15].

Cost of Generating Hydrogen			
With electricity at ($/MWh)	Cost of hydrogen ($/kg)	Cost of hydrogen as gasoline ($/gal)	Cost per mile hydrogen ¢/mile
20	1.11	1.05	2.81
30	1.66	1.58	4.21
40	2.22	2.11	5.62
50	2.77	2.63	7.01
80	4.44	4.22	11.25

Table 17 Hydrogen - Cost of Production

---

[15] For comparison purposes, the cost of driving a 30 mpg car with gasoline at $3/gal is 10¢.

Today, the average price for electricity for the residential market in the US is 10.6 ¢/kWh or $106/MWh. At those prices, it is not economical to generate hydrogen as a liquid fuel.

In addition to the price, there is the matter of practicality. The hypothetical gas station selling 15,000 g/d, consumes approximately 34.7 MWh (working 24 hours a day) for hydrogen generation. To produce sufficient hydrogen to substitute for all the gasoline consumed in the US, even assuming a 25% efficiency improvement in a car propelled with a fuel cell, would require generating 2.25 times more electricity than what it is produced in the US today, going from 4.1 TWh to 13.2 TWh[16].

Since most of the hydrogen is produced from natural gas, the cost of the electrolyzers, adapted from older systems, is estimated in about $750,000/MW. There are small units in the market today at much lower prices, suggesting that the price of the electrolyzers could decrease to $250,000/MW. At $0.5 million per MW, a hydrogen dispensing station with enough storage for three days of selling 4,000 kg/day (about 4,000 gallons of gasoline) would cost about $7-8 million.

Other ideas

Fossil fuels are the densest form of chemical energy by far, and it seems silly to give up that advantage just because the Earth is warming. Maybe we can recycle the carbon. Plants, as a matter of fact, have been doing it for millions of years, capturing CO_2 from the atmosphere and producing burnable products (or products transformable into liquid fuels like alcohol or diesel). Algae in the oceans have been doing it for about 3 billion years, but we do not yet know how to replicate the work of microscopic algae or do it economically.

---

[16] 6.4 k/p d /0.82 (assumed density) gives 7.8 l/p d or 2 g/p d;. using 1 gallon of gasoline is equal to 1 kg of hydrogen, and the requirement of 55.5 kWhr/kg of hydrogen gives 2 kg/p d x 55 kWhr/kg x 320 million people x 365 days/year = 1.28 x 10^{16} watts/year; US generation in 2014 was 4.1 PW (10^{15} w/y)

Conclusion

Not surprisingly, it seems that in a mere 32 pages I was able to arrive at the same conclusion that most government officials and top executives of the oil industry and public utilities have reached. Renewable energy is wishful thinking. If some of you are concerned about CO_2, let's install nuclear plants. The best alternative appears to be continuation of the same.

It is very difficult. Do we rise to the challenge or gamble that there is more oil and that global warming is a temporary natural phenomenon that will correct itself?

I know that there is plenty of research going on. Every week there are articles that such and such university found ways of capturing more photons in photovoltaic panels with new materials or that an artificial enzyme seems to be able to break down cellulose or that the photochemical reaction of substance A shows great potential for generating hydrogen. These and other possibilities need to be studied fully. The future without energy is dark (pun intended!). We might be able to start with the best practices, but systematic research, while the politicians get their act together, costing several billion dollars a year would be money well spent.

I am aware that alternative renewable energies are at a great disadvantage compared to fossil fuels. However, those are the tools we have today. It might be possible to improve the situation somewhat with both practical and applied research, seeking solutions that can be scaled to be implemented worldwide and reducing costs. I believe that there are many dedicated students, professors, PhDs and engineers working non-stop to find solutions or alternatives and that what we have is the best those minds have come up with. There are many interesting ideas that have potential, but many might not have gotten funding because they might not have been competitive with the current price off fossil fuels.

Chapter 8 Cost Estimates

For the next three chapters, I am going to narrow the discussion to the US, even though arresting the repercussions of climate change requires the whole world to participate. If the US were to stop generating CO_2 within 10 years, it would not solve the problem if the rest of the world continued to emit some 28 GTY of CO_2. The feared impact of global warming would just materialize a bit later. In the final chapters, I will expand the scope to include the rest of the world. Leaders always must go first!

Rather than assuming a major breakthrough through research or a substantial improvement in efficiency or reductions in price, the cost estimates are based on today's technologies and current reported prices, whenever possible. I am aware that the current technologies, as discussed above, have substantial limitations, but unfortunately that is all we have.

Writing this chapter was very difficult. It needed to sound realistic, not like science fiction. The topic is full of uncertainties. We do not even know which path to take, except that: (i) some alternatives only work during the day; (ii) others are intermittent; (iii) there are no easy ways to store energy; (iv) hydrogen has a lot of problems, and; (v) nuclear energy is still the unwanted black sheep. We know that economies of scale do not work very well for diluted solar energy and honestly, we have to wonder if some of the alternatives will scale up to the needed level, with millions of devices installed around the world, or if the prices for mass production will drop sufficiently to make them affordable to everyone and not only to the rich in developed countries. We do not even know how inflation or the cost of the energy needed to fabricate the new devices will affect the prices of the new systems. For some items, we do not know if there are enough rare earth materials to fabricate some of the needed components or enough lithium for billions of batteries.

The lack of precision should not discourage progress. The numbers provide an order of magnitude, and by manipulating assumptions, a reasonable range of cost estimate can be secured. Most projects start with very rough estimates and the numbers get refined as you delve further; some assumptions get confirmed and some critical aspects recognized. The numbers are further refined with detailed engineering data, and then the engineering firms bidding for the contract further refine the numbers. Given the scope of the project, nationwide over a period of 30-50 years, the best we can hope for today is an order of magnitude figure.

A good example of the uncertainties is storage using batteries. The lifespan of batteries depends on the severity of the discharge, and with deep discharge cycles, batteries last only about one year. A typical deep cycle 100 Ah battery costs about $250, can hold at 50% discharge about 600 watts and at this discharge rate will last about 5 – 7 years. The average electricity consumption in California (due in part to its nice weather!) is about 750 kWh/month, or 25 kWh/day. To store enough energy for a three-day cloudy period (75 kWh) requires 125 batteries, which cost $31,250. It sounds bad, but doable.

But the number gets bigger if you multiply it by 100 million houses in the US. The cost is just astronomical: $3.1 trillion. That is 20% of the US GDP. If the batteries last five years, it is 4% of the GDP every year. That is $600 billion per year, or about one and one half times the cost of oil imports in the worst year. Since the US generates about 20% of the world's electricity, the price tag for the world would be $15 trillion.

There are some 300 million automotive batteries in the US. We need to produce some 12.5 billion more batteries in the US alone, or 75 billion more for the world. Is the cost going to go down or up? The batteries are already mass produced, with recycled materials, and there is plenty of competition, so the price is likely to remain about the same.

There are few, still in the testing/development stage, industrial flow batteries. Intense research on batteries has been going on for more than 100 years. It might be worthwhile to remember that Henry Ford originally wanted to make electric cars. What happens if electric cars become popular? How many more batteries will we need? Is there enough lithium/lead for 25 or 75 billion batteries?

It's a good thing I'm not advocating the use of batteries to store electricity. The uncertainty is huge, and honestly, a recurring cost of $0.6 trillion per year to become a carbonless society is very high for us in the US but unaffordable to poorer countries.

Investment

The cost estimate is based on ball park figures of the cost of renewable alternatives as presented in Table No. 5 and repeated here. Although nuclear energy is not a renewable source, it is included because it does not generate CO_2. The table also presents the typical availability, the fuel cost for biomass or nuclear options; the capital cost, which is the assumed investment divided by the total energy produced during the life of the equipment, assuming the equipment lasts 20 years, and; the monomic cost, which is the sum of the capital cost and fuel. A supposedly better metric preferred by the DoE is the levelized cost of energy, but that requires assumptions about inflation, interest rates, tax rates, and even maintenance. Bankers prefer the rate of return on an investment, but that requires even more assumptions.

Source	Efficiency	Cost of Fuel ($/MWh)	Investment ($/kw)	Availability (%)	Capital Cost ($/MWh)	Monomic Cost ($/MWh)
Wind Turbines			1,000	25	22.83	22.83
Photovoltaic			2,500	16	89.18	89.18
Solar Thermal High Temp.	32%		7,600	41	105.80	105.80
Solar Thermal Medium Temp. *	16%		4,000	25	91.32	91.32
Biomass	28%	36.57	1,500	75	11.42	47.98
Nuclear	36%	1.20	10,000	95	60.08	61.28

Table 5 Cost of Generating Electricity (repeat)

I have eliminated hydroelectricity, geothermal and ocean energies from the cost analysis because they are site specific, and I will not be considering biomass because it does not scale to the needed level. That limits the alternatives to nuclear, wind, photovoltaic panels and two alternatives for solar thermal (high and medium temperature).

The cost of a PV panel is in the range of $0.60 - $1.00 per watt, if bought piecemeal. The installed cost, including a small inverter, easily jumps to $2.50 to $3.00 per watt. Yes, the estimate is crude, but the estimate of availability is even cruder. A fixed, horizontal PV panel can produce about 1,200 Wh/y per peak watt installed in New York City, or about 1,500 Wh/y in northern California or about 1,800 Wh/y in southern California, Arizona or Nevada, corresponding to an availability in the range of 13.7% to 20.5% (the output is the same, irrespective of the efficiencies, but the area changes with efficiency). The chosen 16% availability reflects a weighted average with population density. PV panels with a nominal capacity of 1 kW peak would be able to produce 1400 kWh/year (16% x 8760 hrs/year). The calculated capital cost, in dollars per MWh, assuming a life of 20 years, would be $89.18/MWh. The range of the estimate is between $69/MWh and $125/MWh.

The cost of wind turbines is about $1 million per MW installed. The cost varies somewhat with the size of the wind farm, its remoteness and the cost of the high tension line needed to deliver the electricity to the grid. Offshore units cost twice as much. The availability depends on the site and is likely to increase in the future since the most promising sites have already been taken. Wind is the least expensive option, but it is also the least reliable.

Nuclear energy is a combination of guesstimates found on the Internet, averaging about $10,000/kW. The US has not built a new nuclear plant in 25 years, and the limited outdated information is likely to be non-compliant with new codes anyway.

Solar thermal high temperature refers to a plant like Andasol in Spain that stores energy in molten salts and can therefore operate more hours a day with a large collector field. Andasol is a 50 MW plant, built in 2009 with an availability of 41% which cost about $380 million.

Finally, solar thermal medium temperature (about 150°- 180°C) refers to a novel approach with fixed collectors, shaped as a relatively flat portion of a spiral, which stores thermal energy in a pressurized hot water concrete tank and can be used to operate as a peaking plant some six hours a day, with a cost estimate of about $1 million for a 250 kW organic Rankine engine. It is the only technology that has not been demonstrated. The problem is not engineering, but cost. It is included because the advantage of having distributed generation at a community level, with the possibility of operating it as a peaking plant, is attractive. In poorer countries, its simplicity might allow communities to have steady electricity during the day and part of the night.

Demand

The second portion of the problem is how much electricity needs to be generated by alternative sources to achieve the goal of becoming a carbonless society. As before, there are uncertainties because there is the expectation of reduced future need due to improved efficiencies and greater energy price awareness which would reduce wasteful consumption, coupled with potentially higher demand due to population growth and an improved standard of living. For the purpose of this exercise, we will assume that the possible growth is cancelled by the potential reduction.

Table No. 2, repeated here, shows the per capita consumption in the US and the translation of the primary sources to satisfy the main uses of energy.

For transportation, we will assume that we will be using mostly hydrogen, either directly as a fuel or in fuel cells. In the previous chapter, we indicated that it requires about 55 kWh/kg of hydrogen and that one kg of hydrogen has about the same energy content (95%) as one gallon of gasoline. The US needs about 5.65 kg of oil equivalent/p d which translates into 1.65 gal of oil equivalent/day (assuming a density of 0.9 kg/l) or 193 billions of gallons of oil per year, and one barrel of oil produces about 31 gallons of gasoline and diesel. The US consumes about 142 billion gallons of gasoline per year. The electricity needed to produce the hydrogen to substitute for this fuel is 7.8 PWh/y, or about twice the current production of electricity. This number can be reduced slightly (20%), given that fuel cells are more efficient than internal combustion engines. These means the US needs about 6.2 PWh/y for transportation.

U.S.A. Per Capita Energy Consumption					
	(kg of oil equivalent per day)				
Source	Mill T/y	kg/p d	kg CO2/p d	Use	kg/p d
Oil	836.1	7.16	24.74	Transportation	5.65
Natural Gas	695.3	5.95	18.01	Other uses of Oil	1.79
Coal	453.4	3.88	15.66	Heating	1.19
Nuclear	189.8	1.63		Other uses of NG	2.38
Hydroelectric	59.1	0.51		Electricity	8.67
Renewables	65.0	0.56			
Total	2,298.7	19.68	58.41		19.68

Table 2 U.S. Per Capita Energy Consumption (repeat)

Just because I provided a number for transportation, it does not mean that the problem is solved. As expected there are many uncertainties here too. As discussed before, hydrogen might not be the solution for all the transportation problems. It is unlikely that hydrogen stations could be placed in high density areas ("not in my backyard" syndrome), at least until everyone feels safe living near a potentially dangerous station. It is likely that electric vehicles "EV" will be more prevalent in high density areas. City residents usually drive fewer miles, so the range of electric vehicles will probably not be a big problem for them. From the point of view of the demand for electricity, there could be some savings. Interstate truck driving faces another hurdle. The thickness of the walls (and consequently the weight) of the storage tank is proportional to the size of the tank, and 18 wheelers can carry up to 100 gallons. The added dead weight is bad for the trucking business, and so is stopping more often to fill smaller tanks. It is easy to imagine a substantial increase in rail transportation (which could be easily electrified), which again reduces the need for hydrogen. Furthermore, neither hydrogen nor electricity are today possible solutions for air travel or ocean freight.

Whenever possible, most of the other uses of oil, natural gas and heating will also need to be substituted for by renewable energy. Some changes might be easy to implement, while others (e.g. plastics) might linger a long time, while we figure out ways to make substitutes. However, as the goal is to be a carbonless society and the program is expected to take 30 or more years to complete, we will be assuming that all will eventually be replaced, using the same conversion factor of 55 kWh as for the generation of hydrogen, resulting in an additional 7.4 PWh/y. I know that I pulled that number out of a hat, but the myriad of products prevent a better assessment.

All the nuclear energy, all the hydro and 50% (assumed) of the renewable energy is used to produce electricity, representing 27% of the total generation. If total generation in the US is 4.1 PWh/y, the generation by fossil fuels is 3 PWh/y. The demand is still 4.1 PWh/y, but 1.1 PWh/y is already carbonless generation.

The total demand for renewable energy to produce the electricity that a carbonless American society needs is the sum of the needs for transportation, other uses of oil and natural gas, electricity and gas for heating, totaling 17.7 PWh/y, of which carbonless generation is 1.1 PWh/y, for a net need of renewable generation of 16.6 PWh/y, which is about four times the current generation of electricity. As mentioned before, we assume the number is static, with growth cancelled by reductions and we will use 16.6 PWh/y as the net future demand of electricity.

Some of you might think that I simplified matters tremendously and that the conclusions are going to be off (garbage in, garbage out!). I assumed, maybe too quickly, that all uses of oil could be easily substituted for. How naïve! Some of them need carbon! We might be able to generate ammonia without it, but urea needs a CO_2 molecule. The building block of plastics is ethylene, and again, that contains carbon atoms. How do we produce benzene, or toluene, or tetra-ethylene glycol without carbon? Maybe we can use biomass to produce bio-plastics! I know the oversimplification is not applicable, and this is one of the reasons I am calling for massive amounts of funding for research. Since the cost is going to be huge, I believe that many will agree for the need to become less wasteful and therefore total demand for plastic bags or polystyrene cups will have to be less and other alternatives need to be found. Therefore, at the risk of sounding stupid, I will continue to use this simple estimate as the ceiling for the future demand of electricity.

Generation Mix

A third uncertainty is the possible generation mix. There are places with little solar irradiation or wind. The high cost of storage and the fact that PV only works during the day and the wind is unpredictable requires limiting their participation to manageable levels. In addition, there is a need for base load plants to serve industries that have around-the-clock production and cannot be stopped and restarted. To the despair of some, we have to accept that nuclear energy must play a stabilizing role.

The chosen renewable energy mix will have to be optimized to the locally available resources, but the high tension grid will need to be expanded dramatically to move more energy from one place where there might be abundant resources to other places with fewer resources or to places where there is cloud cover.

Fortunately, a good portion of the energy is going to be needed for hydrogen production, which, given the inherent difficulties of transporting hydrogen, said production should be distributed as much as possible (it is easier to move electricity than hydrogen). The production of hydrogen could be used as a shock absorber to provide stability to the grid by reducing or increasing production depending on the availability of electricity. Controls could sense change in the line frequency or monitor the prevailing cost of electricity and only take energy from the grid when the price is below a set threshold, adjusting the consumption accordingly. It resembles the much touted smart grid, designed for heavy industrial consumers or hydrogen generators.

The flexibility provided by the market's adjustable demand will allow renewable energy to provide the kind of electricity we are accustomed to. Using hydrogen to operate thermal plants during periods of persistent cloud cover might sound appealing, but the efficiencies are very low and it would be almost impossible to store sufficient hydrogen to operate large thermoelectric plants.

Rather than attempting to predict the optimal mix for the whole country, several possible alternatives are presented in Table No. 18. Interestingly enough, the costs are similar, although the size of the needed installed capacity differs between alternatives.

Possible Generation Mix and Cost

Energy Source	Investment ($'000/MW)	Availability	Monomic Cost ($/MW)	Flat	Generation Mix Low Nuclear	Moderate Nuclear	High Nuclear
Nuclear	10,000	95%	48.07	20%	5%	15%	25%
Wind	1,000	25%	18.26	20%	30%	25%	25%
PV	2,500	17%	67.15	20%	20%	20%	15%
Thermal - High	7,600	41%	84.64	20%	30%	25%	25%
Thermal - Low	4,000	16%	114.16	20%	15%	15%	10%
Average Cost ($/MW)				5,020	4,180	4,750	5,425
Average Availabilty				38.8%	30.4%	36.6%	44.4%
Monomic Cost ($/MWh)				66.46	63.83	63.49	59.23
Installed Capacity (GW)				7,437	7,766	7,355	6,405
Estimated Cost ($ Trillion)				27.6	26.5	26.3	24.6

Table 18 Possible Generation Mix

The installed capacity needed to provide 16.6 PWh of electricity ranges from 6.4 TWh to 7.7 TWh (six to seven times the existing capacity today), costing between $24.6 and $27.6 trillion (about 1.6 to 1.8 times the current GDP). It will require installing between 100 and 500 new one GW nuclear power plants, about two million one MW wind turbines and some 25,000 50 MW solar thermal plants like Andasol, among others.

It appears that I have sidestepped the problem of storage. The high nuclear option coupled with high temperature solar thermal should provide base load capabilities with low temperature solar thermal facilities acting like peaking plants. But the main flexibility is provided by having the hydrogen generation expand or shrink to absorb the peaks and valleys of generation. A station with low hydrogen inventory will have to pay higher rates for electricity, so the free market provides the needed flexibility.

Additional Requirements

We are not yet through with uncertainties. The high voltage power grid will have to be expanded drastically, first to accept the new electricity that is going to be generated, but also to move it from region to region, with the Southwest producing most of the solar thermal energy, and Texas, California, and some Midwest states producing most of the wind energy.

Transmission lines represent about 30% of the replacement cost of today's electric sector, or about $1 trillion dollars, for about 1 TW installed capacity. If the proportion were to hold, the required investment would be $4 to 7 trillion. That is a lot of money for such a superficial estimate, but a more reliable figure requires substantial details that are not available. We know that generation is likely to be more distributed, requiring fewer transmission lines, but we also know that it would make sense to have distributed generation of hydrogen (it is cheaper and more efficient to move electrons than molecules). We also know that we need a stronger and more redundant system to shift electricity back and forth, supplying zones with cloud cover with electricity generated elsewhere. The need to accept a large portion of distributed intermittent generation, couple with distributed hydrogen generation reduces the need of huge expansion of the transmission system and the lower range of the estimates is selected and reduced to $3 trillion by absorbing the existing infrastructure.

The uncertainties of transmission lines pale in comparison to the uncertainties of hydrogen generation. Most of the hydrogen is produced from natural gas, and while electrolysis is well known there are not many installations to draw data from. There is the cost of the electrolyzer, plus the compression equipment (to 5,000 psi or even higher), plus strong storage tanks whose wall thickness depends on the size of the tank, plus the transportation of the product to dispensing stations plus the cost of the new dispensing stations. Most gas stations are privately owned, but since the cost of the new hydrogen stations is going to be much higher than that of a regular gas station, it would be wise to include the estimate in the total cost of the carbon-free system.

We need 6.2 PWh to generate some 140 billion kg of hydrogen. Using 55 kWh/kg of H_2, a 1 MW electrolyzer would be able to produce about 18 kg of hydrogen per hour, or about 360 kg of H_2 per day (assuming 20 hours of operation to shift in and out, depending on the availability of electricity). A typical gas station sells about 4,000 gallons of gasoline per day, and therefore we will need 11, 1 MW electrolyzers. Using $0.5 million/MW, and also $0.5 million each for compressors, tank, electrical, control and dispensing units, the total cost per gas station is about $8 million. To produce the amount of hydrogen needed to substitute for all the gasoline, we would need about 96,000 hydrogen stations (there are about 125,000 gas stations in the US), so the cost is likely to be about $800 billion.

If we wanted to have thermoelectric plants burning H_2 for cloudy days, investing another $800 billion would generate about a third of the electricity needed today or about 8% of the energy needed in the future (which includes the need to generate hydrogen). It should be noted that the scheme is highly inefficient. It requires 55 kWh to produce 1 kg of hydrogen, but burning the hydrogen in a peaking plant only produces about 10 kWh. It only makes sense if there is an oversupply of electricity due to excessive wind or photovoltaic output.

For simplicity and round numbers, we will use $2 trillion for the cost of producing and dispensing hydrogen for transportation and for backup or peaking thermoelectric plants.

Some claim that there is also the possibility of using the EV and fuel cell ("FC") cars as storage devices. In theory, an EV connected to a charging station could provide energy back to the grid if the price was attractive, or an FC car could generate electricity and send it to the grid. While technically possible, it would require additional investment in inverters, synchronizers, etc., plus the willingness of people to have their battery or hydrogen tank drained during cloudy days, hoping that the sun will shine tomorrow and they can go to work with what was left in their battery or hydrogen tank. It might work, but the incentive (i.e. the price differential) would have to be large for people to contribute.

While it sounds easy, the devil is in the details. If the number of electric cars is much higher than assumed and rail transportation reduces the need for trucking, the need for hydrogen and therefore the flexibility provided by adjusting electricity demand with hydrogen generation could be smaller and the penetration of intermittent sources (i.e. PV and wind) would have to be curtailed.

In summary, the total cost to replace all fossil fuel with renewable energy is estimated to be $30 trillion ($25 trillion for generation, $3 trillion for transmission and $2 trillion for hydrogen generation and dispensing). The cost is about two times the present GDP. It is a fixed cost based on today's assumptions and assumes no inflation, no shortages or speculation on the prices of strategic materials. Since the base investment requirements are gathered from many sources, it is safe to assume that they include interest during construction, taxes and even engineering and procurement costs. It also does not assume price reductions due to economies of scale or mass production.

Government Participation

The astronomical cost of $30 trillion is well beyond the financing capabilities of the private companies in the electric sector and would prevent the private sector from implementing the change in a timely fashion, not withstanding all the free market rhetoric[17]. The book value of the electric sector is about $3 trillion,

---

[17] The oil companies might not be able to raise funds either because the market would be concerned about their repayment capability, given the push to eliminate CO_2 emissions.

126

with a debt equity ratio of about 50/50 and total sales of about $0.4 trillion. The sector simply cannot raise some $15 trillion in debt plus issue $15 trillion in shares, or about $0.5 trillion per year in debt and $0.5 million in equity to finance the conversion to a carbonless society in 30 years unless the proceeds of the carbon tax or another scheme implemented by the government are given for free to the sector. The interest of the debt plus the return on equity for a $30 trillion investment would need prices to skyrocket beyond the means of 99% of the users. The same would be true if the total cost were only $15 trillion.

Government involvement is consistent, with its mandate of fostering the well-being of its citizens. Historically, many countries and even many municipalities in the US took it upon themselves to be responsible for the generation, transmission and distribution of electricity because it is a natural monopoly subject to abuse and because the private sector lacked the funds to implement it at the speed the government might have wanted for its citizens to enjoy the benefits of electrification. In the US, the government implemented the Rural Electrification Act, providing soft loans to municipalities; the US Corp of Engineers constructed most of the large dams, and the federally owned Tennessee Valley Authority is the largest utility in the US. In almost all countries, the government has taken on the responsibility of building and maintaining public roads, most of them free. Initially some bridges and later some roads have used toll booths to pay for at least part of their construction. Some new roads or express lines on existing roads have been built by the private sector, which collects tolls for their use, but most road construction and maintenance is still a government responsibility.

Becoming a carbonless society is a task that only the government can achieve. Only the government can muster the resources to implement a rural electrification goal, build the interstate network, provide education for all children, implement the Marshall Plan or achieve the goal of reaching the Moon within a decade. It is the only entity that can raise the needed funds, organize and prioritize their implementation and mobilize the population into the crusade to save the planet. It might be federally mandated and coordinated, but should be implemented by the states or even municipalities, taking into consideration the availability of solar resources and population densities, seeking the best alternative at the state level that is consistent with the overall goal.

Furthermore, it is the only entity that can shoulder the responsibility of building new nuclear plants, implementing a solution to nuclear waste and providing the level of security needed, or assume responsibility for litigation likely to emanate from privately owned electric companies fighting for their obsolete monopolies. I am not suggesting that the government steamroll all opposition, but emergency situations require emergency measures.

Most of the opposition to the fight against global warming is being backed by entities trying to preserve the status quo, protecting their markets and profits rather than the planet. There is no way to placate their voices and opposition. There is no way to enlist their support, unless they are given the keys to the Treasury. The public is going to have to pay for the investment in renewable energy and bear all the inconveniences and sacrifices required of them, and the government has an obligation to administer those funds to maintain and improve the welfare of all citizens, not only the companies that have benefited from agreements or laws written before we became aware that burning fossil fuels had negative consequences.

CO_2 Emission by Typical Units

The best way to arrest the relentless increase of CO_2 in the atmosphere is by making it more expensive. It is very difficult to quantify the cost of "externalities", those factors that we have taken for granted for centuries and that are now showing unforeseen effects. The cost of one more part per million of CO_2 in the atmosphere to our society is very difficult to quantify. Is it a cost only to the present generation or should we include future generations? Fortunately, we do not need to answer that. We know that emitting CO_2 is bad, and we want to avoid emitting more. We know that fossil fuels emit CO_2, and we can put a price on those emissions. What price? Whatever is necessary to finance the conversion to a carbonless society? Based on the estimate, we need about $30 trillion. If the program is to be implemented over ten years, we would need to raise $3 trillion per year to finance it; if over 30 years, about $1 trillion per year, and if over 60 year $0.5 trillion per year.

Eventually, the funds are going to come from all individuals. There are two possible alternative ways of doing it: (a) the fee can be affixed to the primary source of energy (i.e. oil, natural gas and coal) and be collected by the entities selling them (refineries, mines or gas wells), or; (b) we can calculate the price per gallon of gasoline, kWh or MBTU and collect it directly from the customer at the gas station or through the public utility providing the gas and electricity. The second alterative has more collection points, but is more transparent to the public and would have more reinforcing impact. Besides, the utilities and gas stations are already acting as collecting agents.

The second alternative requires expressing the cost of CO_2 per typical consumption unit. The public buys gasoline by the gallon, electricity per kWh and industry buys natural gas by MBTU. The amount of CO_2 emitted by each one of the primary sources of energy is given in Table No. 19.

The stoichiometric figures assume oil is C_8H_{16}, coal does not have impurities and natural gas is methane. The ratio reflects the weight of product per ton of oil equivalent. The line labeled Consumption MTOE is the consumption in 2014 in million tons of oil equivalent and the line labeled CO2 emission Mtons reflects the amount of CO_2 emitted by each source. The line labeled Use means the utilization of each source in typical units in 2014 and finally the lines named Emission and Units reflect the emission of CO_2 by each source expressed in typical units used.

CO2 in Typical Units			
Source	Coal	Oil (C8H16)	NG
Stoichiometric	3.667	3.143	2.750
Ratio	1.500	1.000	0.809
	ton coal/toe	tons oil/toe	MBTU/toe
t CO2/Ton oe	5.500	3.143	2.225
Consumption MTOE	453.4	836.1	695.3
CO2 emission Mtons	2,493.7	2,627.7	1,546.9
Use	909	142	22.8
Units	million tons	billion gallons	Tcuft
Emission	2.74	18.51	67.85
Units	tons CO2/ton	kg CO2/gal	kg CO2/MBTU

Table 19 CO2 Emissions

Oil is used to produce fuels (gasoline, diesel, heating oil and jet fuel) and for other purposes (petrochemicals, lubricants, paving, etc). The output changes from refinery to refinery, due to market conditions or to cracking severity in response to seasonal demand. The approach taken previously of using 75% of the oil for transportation is a good approximation, and therefore the CO_2 emitted per gallon of gasoline will be adjusted to 13.87 kg CO_2 per gallon. For the purpose of pricing CO_2 for other uses, it might not be very accurate or representative. However, management and engineering skills at refineries are excellent, and properly supervised by DoE personnel, a scheme could be worked out to translate the generation of CO_2/gallon of gasoline to equivalent ratios for ethylene, polypropylene, benzene, high temperature lubrication oil or even asphalt. While higher gasoline and electricity prices would make a direct dent on the emission of CO_2 from those products, failing to simultaneously increase the price of plastic bags or plastic bottles would hinder a decrease in demand for them.

Natural gas is usually traded in MBTUs (or MMBTUs for some) although for residential use, a smaller unit, the therm, equivalent to 0.1 MBTU, is used (1,000 cubic feet of natural gas at STP - standard temperature and pressure –contain about 1 MBTU). It is common in the industry to have different rates for industrial, commercial and residential customers, with the last usually paying the highest rates because it is less efficient to distribute to residential areas than a big pipeline to a factory located nearby (the amount invested per client is larger). For the purpose of the carbon fee, the industry can keep their current practices, but the fee should be the same across the board, since one kg of CO_2 emitted by industrial customers is just as bad as one emitted by residential clients. The carbon fee for natural gas used for electricity would be priced separately in electricity.

Finally, electricity is a bit more complicated. To calculate it, we used all the CO_2 emissions from coal and 40% of the CO_2 emissions from natural gas (assigned to electricity), divided by the total production of electricity from fossil fuels (nuclear energy and hydroelectricity already in place do not emit CO_2). The total emission for the generation of electricity is calculated as 3,112.4 million tons of CO_2 per year, for the generation of 3.0TWh per year, or 1.039 kg CO_2/kWh.

Not all the utilities emit the same amount of CO_2 per kWh. The amount of CO_2 emitted by each power plant depends on the fuel they use and the efficiency of the plant, with newer units being more efficient than older ones. The industry and regulators have reporting mechanisms in place, and the amount of CO_2 emitted by each plant is known.

There are two options: (a) have a fixed fee per kWh, irrespective of how it is generated, or; (b) have each utility calculate their own emission coefficient by dividing the total amount of CO_2 emitted during the period by the generation during the same period. The first alternative might be easier to implement, but the second alternative places more pressure on the utilities to more quickly change their generation mix to less emitting facilities, otherwise their customers will face higher electricity prices and thus reduce their consumption faster. Different fees will result in the same average.

With the caveats and simplification discussed above, the calculated emission per typical unit of primary energy was translated into emissions per units of energy that the customers buy daily. To summarize, the emissions per typical user unit are: 13.87 kg CO_2/gallon of liquid fuel (or refinery product), 67.85 kg of CO_2/MBTU for natural gas, and 1.04 kg CO_2/kWh consumed.

CO_2 Pricing

The purpose of collecting a carbon fee per kg of CO_2 emitted is to finance the conversion of the system to create a carbonless society. We now have an estimated cost of $30 trillion, and we also know that the current emissions by the US are 6.7–7 GTY of CO_2. The missing variable is the time frame.

The program could be carried out over 10 years or 100 years. It was discussed before that a linear reduction in 80 years would have CO_2 reach 478 ppm by the year 2100 and linear reduction in 50 years would result in CO_2 reaching 456 ppm by the same year. Repeating the exercise for 30 years would limit the concentration of CO_2 to 440.4 ppm. It seems a possible goal. With 30 years as a goal, the program needs to collect about $1 trillion dollars in carbon fees per year.

The carbon fee is the division of $1 trillion by the amount of CO_2 emitted (calculated to be 6,668.3 million tons in 2014), or a price per ton of CO_2 of $150. Expressed in typical units, the carbon fee needed to raise $1 trillion per year is $2.07 per gallon of gasoline or fossil fuel; $10.17 per MBTU for natural gas and 15.6¢/kWh for electricity.

The numbers are huge and would obviously have a negative impact on the economy, but fortunately, all the money would be invested in the country and thus also act as a stimulus. The number could in theory be reduced in several ways:

- improved efficiency – regulations to improve the efficiency of transportation and home devices would lower the need for alternative renewable sources reducing the cost;

- reduce unnecessary consumption – implementing regulations such that cars automatically turn their engine off at stop lights; install switch sensors that automatically turn off the lights when rooms are empty, and; a mandate to eliminate parasite losses, would lower the cost of the program;
- expand funding for research into renewable energy ten or fifty times – it is ridiculous to embark on a $30 trillion program spending about $100 million/y in research;
- expand the time frame – if the program were carried out over 40 years, the concentration in the atmosphere would reach 450 ppm, but the investment needed will be reduced $0.75 trillion per year and the carbon fee will decrease to $112.5 per ton of CO_2, with the fee per typical unit reduced to 75% of the figure given above. It would be tempting to expand it further to 100 or 200 years; however, there might not be a habitable planet by then;
- encourage/facilitate changes of habits – provide better public transportation, discourage the spread of suburbia, encourage local production/consumption; incentivize smaller cars, etc.

Table No. 20 shows the carbon fee per typical unit for several possible time and extent scenarios. The table shows the needed carbon fee in typical units according to various scenarios, varying the time needed to implement the program and the desired level of energy relative to the level used in the year 2014. A reduction to the level of 50-60% would put US energy consumption at the current level of consumption in many European countries and would be the result of having a higher cost of energy. The new prices of energy including the carbon fee would still be below what many other developed countries have.

The temptation to expand the time frame or optimistically assume less future energy needs to lower the carbon fee should be rationally discussed to avoid pretending that we are doing something while we are really continuing to do business as usual.

Carbon Fee - Time and Extent							
Energy (%)		100	90	80	70	60	50
Investmet $T		30	27	24	21	18	15
				Program in 30 years			
CO_2 ($/ton)		150	135	120	105	90	75
$/Gallon	0.01382	2.07	1.87	1.66	1.45	1.24	1.04
$/MBTU	0.06785	10.18	9.16	8.14	7.12	6.11	5.09
$/kWh	0.00104	0.156	0.140	0.125	0.109	0.094	0.078
				Program in 40 years			
CO_2 ($/ton)		112.5	101.25	90	78.75	67.5	56.25
$/Gallon	0.01037	1.55	1.40	1.24	1.09	0.93	0.78
$/MBTU	0.05089	7.63	6.87	6.11	5.34	4.58	3.82
$/kWh	0.00078	0.117	0.105	0.094	0.082	0.070	0.059
				Program in 50 years			
CO_2 ($/ton)		90	81	72	63	54	45
$/Gallon	0.00829	1.24	1.12	1.00	0.87	0.75	0.62
$/MBTU	0.04071	6.11	5.50	4.89	4.27	3.66	3.05
$/kWh	0.00062	0.094	0.084	0.075	0.066	0.056	0.047

Table 20 Carbon Fee Scenarios

Future Adjustments

So far the picture is static, constant with no inflation or break-throughs. Reality is going to be messier. The program would start slowly after much deliberation and solving of land issues, allocation of funds to municipalities, cities and states, solving legal issues and compensation for using existing infrastructure or rights of way, normal inertia, building or ramping up production of needed components, training builders and even graduating engineers. Inflation is going to be unavoidable. Critical raw materials might not be available in the quantities demanded, plus prices for everything, including labor, would start reflecting the carbon fee paid for its manufacture or transport.

Furthermore, the success of the program (the reduction of emitted CO_2) would lower the base for collecting the carbon fee. The need of funds might be the same, but now the amounts collected would be less because the public would have shifted to more efficient cars or public transportation or walking, the houses would have been properly insulated and people have become more wary of wasting expensive energy. Despite the success, there is no alternative but to increase the carbon fee, making it somewhat regressive because wealthier people can afford electric or fuel cell cars, but poor people are still going to be driving now inexpensive gas guzzlers.

Maybe the painful alternative of starting with a high carbon fee and keeping it constant for a long time, with the government stashing funds for future investment would provide less shock to the economy than continuously adjusting the carbon fee. It might be tempting to change it to an energy fee rather than a carbon fee, and while demand for energy might also decrease, the energy fee would be applicable to hydrogen and electricity generated by wind or solar energy, but that removes a strong motivation to eliminate carbon emissions quicker. If driving to work is going to cost $20 with a gasoline car, using public transportation only $4 and a fuel cell car $7, it makes sense to donate your SUV to your favorite charity. Hopefully, the reduction of CO_2 emissions might start slowly, but finish quickly.

It might be possible to have breakthroughs, with cheaper materials, higher efficiencies, improved manufacturing, or genuine technological miracles that might reduce the cost or the need. If it happens, we will have learned to respect the environment and sustainability and will have some local and renewable energy, and know that when we were challenged, we were brave and took the steps needed at the time.

The world demand estimate could be four to six times the American estimate. The US consumes almost 20% of the primary energy, therefore five times is a reasonable estimate. There are countries with colder climates that might need to dedicate more energy to heating, and there are countries with less area and therefore need less energy for transportation. Furthermore, there are countries with more solar irradiation, so the efficiency of collecting solar energy could be better.

Scaling up

We have a rough cost estimate now, but is it realistic? Does it scale up? Are there enough rare earth materials? The list goes on and on.

Where do we put 500 new nuclear plants (10 per state?) next to large population centers (but not in anyone's back yard) to minimize transmission, but also next to water for cooling? How many tons of sodium are we going to need for cooling the Gen IV nuclear plants? How to we transport it, or do we separate it from salt at the site? Where do we get the uranium? For high temperature solar thermal, where do we get 650 million tons of eutectic salts (sodium-potassium nitrates)? Where do we install 20,000 -30,000 units, each requiring 0.5 km^2 of collectors which will have to be transported by some 900 truck trips, assuming we can pile them 10 units high? How do we put hydrogen stations in high density centers? Safety is critical. Maybe high density cities will have to depend on or use electric cars.

The list of problems goes on and on, and we despair. It is a heck of a lot of a sacrifice with so many uncertainties. Is it worth it?

Watch out! Doubts translate into inaction and inaction into business as usual (i.e. doom or global warming!). We need to solve the problem. Therefore, the suggestion for increasing funds for research some 10, even 100 times should be given priority.

It is so easy to dismiss the idea as simplistic, naïve, uninformed, unrealistic, unreasonable or similar adjectives. I am calling for an effort to gradually eliminate CO_2 in the next 30-50 years, starting in 2020. I am giving some five years for the political forces to fit the square peg on a circular hole, hoping that while bitter discussions are taking place, signals are send to: (i) the DoE to expand tremendously research on the many problems that need to be solved, refine the numbers and start planning, and; (ii) legislators to study the mechanisms and needs to implement the changes.

Chapter 9 Limiting Mechanisms

We have already discussed that:

- the majority of the scientists agree that the burning of fossil fuels and their subsequent production of CO2 is affecting the Earth's climate;
- the Kyoto Protocol to failed to limit, curtail or reduce the emission of greenhouse gases;
- the world produces today some 35 GTY of CO_2 per year, of which the US produces about 7 GTY, representing about 20% of the total emissions by less than 5% of the population and, on average, each American produces some 20 metric tons per year, or about 58 kg per day;
- politicians are reluctant to act while some less developed countries continue burning fossil fuels.

No one knows if there is a "safe" level of CO_2. No one knows for certain if the sea level rise will be fast or slow, or how much or if the permafrost will start releasing methane or if the Gulf Stream might stop. Unfortunately, I do not have the expertise to express an opinion about it. There are plenty of other books describing expected consequences. The limits suggested today are so vague, bordering in the ridiculous. How can we predict the level of CO_2 that will limit the warming to only 2°C? Does that level leave snow on the mountains for summer irrigation?

However, I need to be blunt. We have to face reality. The only way to curtail the emission of CO_2 is by eliminating the burning of fossil fuels. Carbon sequestration is a myth – I will repeat my assertion that "clean coal" is the best example of an oxymoron. There is no way to sequester 35 GTY. First, there is no place to put them, even if it could be compressed at several thousand pounds per square inch of pressure. It would still require hundreds of cubic kilometers of storage space. It would require huge amounts of energy to concentrate and compress. It would require extensive pipelines from the emission site to the storage site and it would only work for concentrated sources of energy, but cars and trucks worldwide will continue emitting some 10 to 15 GTY.

All the talks about curbing emissions to the 1990 level, or reducing emissions 20% by 2020, 25% by 2025 or even the whopping California goal of 33% by 2020 are good intentions, but do not much change the outlook. For a family (or a country) in debt, reducing future borrowings by 20, 25 or 33% of the borrowing during the good times is not going to improve their financial situation much. Drastic measures are necessary – no more borrowing and start paying some of the debt back. During the 90's, the concentration of CO_2 in the atmosphere grew on about 2 ppm per year. Reducing emissions so that the growth is only 1.6 or even 1.33 ppm per year is good, but the effect would be negligible.

Stopping burning fossil fuels is probably going to be a game changer. It could mean good bye to living in the suburbs and commuting to work or to imported flowers from Colombia or asparagus from South America or imported beers. We are going to have to reduce the miles our food is driven, the lighting on the streets, the humming of the air conditioners. Welcome empty freeways, community stores and community activities and work.

This book is written on the assumption that the increasing CO_2 concentration is bad for the planet's health and to explore possible avenues to curtail it. While they might be phrased differently or combined in some fashion, there are only five possible ways to curtail the emissions of CO_2. They are: voluntary measures, regulation, cap and trade, carbon tax and rationing. Some have less political cost, while others might be more effective or fairer.

Voluntary measures

The politicians can inform the American people of the dangers of continuing to burn fossil fuels and ask everybody to voluntarily contribute to reduce the emissions – set thermostats a bit lower in winter, use less air conditioning in summer, walk, car pool, use public transportation. It can be complemented with big pushes for efficiency, weatherization, green buildings, electric vehicles, planting trees, composting, recycling, smart grids, even some in-centives for high mileage vehicles or solar panels, etc. Disruptions would be few. In normal markets, the cost of fuels and electricity should decrease (there is less demand) and everybody would live happily ever after. Renewable energy might not be able to compete with low fuel prices, and therefore its penetration would remain small.

Just like exercise, diets or smoking, few New Year resolutions last more than a couple of weeks, and suddenly we are back to square one, driving 20 miles to save $1 on some beer, the thermostats creep slowly back up to their original settings and we are still burning tons of fuels. With constant reminders and effective PR campaigns, we might be able to achieve a small reduction of emissions and slow down the growth rate.

For the sake of argument, let's assume that, with strong efforts in developed countries, we succeed in reducing the growth rate of emissions from 2% p.a. to 1% p.a. gradually for the next 40 years. Rather than reaching 478 ppm by 2050, we would have made some significant progress and the CO_2 in the atmosphere would be only 469 ppm.

It would cost very little – let's just say that the contributions to National Public Radio will now require government-sponsored reminders of the need to conserve electricity and save fuel. It costs little, but we get very little.

In the last 10-15 years, we have seen the poor results achieved by voluntary contributions. The Kyoto Protocol was signed in 1997, and many European countries embarked on far-reaching efforts to adhere to its dictates. The US did not sign it, but some folks have been concerned and have made small contributions. Granted, it was not a bona fide effort, but the end result is that the growth rate on the concentration of CO_2 was the largest ever experienced, growing on the average 2.214 ppm per year during the decade. We can blame China, India and a bunch of other less developed countries for increasing their emissions disproportionally, but the reality is that China's CO_2 emission per capita is about 6 tons per year, while the United States' is 19.34 tons per year.

It is fair to conclude that voluntary measures are not sufficient to arrest the emissions of CO_2. It is the most ineffective of all the alternatives. It maintains the status quo and the belief that technology will conquer all, robbing future generations of the possibility of enjoying life the way we were fortunate to be able to. It is a dishonest and cowardly approach. Good leaders lead through good and bad times, irrespective of whether they might be hurting their chances for reelection.

Regulations

The government can mandate many efficiency requirements, decouple the profits of utilities from selling more electricity, require building, houses, appliances and automobiles to drastically improve their efficiencies, incentivize conservation, etc.

Some measures could imply changing the construction codes requiring compliance with house orientation, level of insulation, three-pane windows, etc. That could save some 10-20% of the future energy needs. At the federal level, the government could require a Corporate Average Fuel Efficiency (CAFE) of 70-80 mpg or a renewable portfolio mandate of 30% by 2030.

All those measures can and should be implemented, but they are insufficient. The overall mandate should be to become a carbonless society by 2050 or 2060, or in the worst case, by 2070. Only by leading and showing the needed strong resolve can we influence other countries to follow.

Furthermore, you can only regulate what is doable. You cannot regulate that everybody get an electric car by 2020. Not everyone is capable of buying a new car in five years (plus - there are not enough electric cars to be bought and the electricity system cannot charge all the cars). Regulating it with a doable time frame, e.g. by 2050, might work.

Cap and Trade Approach

The cap and trade mechanism is rather simple and was extremely successful in reducing emissions of SOx. The government gave annual permits (to each known plant that emitted SOx) to emit a given amount of a SOx, effectively capping the amount that could be emitted that year. The amount was reduced overtime. Plants could curtail production, install equipment to reduce emissions or buy permits from other plants that had excess permits. The results exceeded optimistic expectations. SOx emissions were reduced faster, deeper and cheaper than other considered mechanisms. Not surprisingly, the success of the cap and trade mechanism for reducing SOx emissions has become the preferred alternative for reducing CO_2 emissions. The cap and trade mechanism suits advocates of free market and those opposing additional government regulations and/or mingling.

Cap and trade is the politically palatable solution in the short run. We let market forces choose the best solution. It is easy to administer. It gives all CO_2 emitters permission to continue emitting next year a fraction of the CO_2 that they emitted the previous year. They either find a way to continue producing at the level they want, reduce production or buy the right to emit more CO_2 from others who were smarter and found a way to reduce their emissions. If the permits are given for free, no one is taxed. We give the permits to big industries that are sophisticated enough to know how much CO_2 they are emitting now, and we can ratchet down the size of the permits slowly and gradually to avoid complaints. Those found emitting more than they are authorized could be fined (hopefully with meaningful fines and not just slaps on the hand).

Market forces will find the best way to reduce emissions. Some utilities might expand their renewable generation, some might shut down coal plants and install combined cycle systems, and some industries might find ways to improve their efficiency or capture waste heat or buy goods from others.

Prices of goods will increase inexorably to compensate the companies for the additional investments made to reduce emissions. The increase might be gradual, but we all recognize that companies have the right to be profitable and compensated for their efforts. If the government charged for the permits, the prices would increase faster, but the government would be blamed for the price of the permit (it would quickly become known as the "carbon tax") and for the needed investments and for the higher market price of the right to emit (which now includes the cost of the permits). The government would take all the blame, without any control as to how (if?) it gets implemented.

It is politically palatable in the short run, because initial reductions are easier to make, but as the level of reductions increases, the costs and investments will grow, awakening protests that will blame government intervention.

It worked fine for SOx because there were only 211 companies emitting SOx (in about 3,500 plants), all of them large and financially sophisticated enough to understand the consequences of non-compliance, and there were several reasonably priced alternatives (shutting down a coal plant and replacing it with a gas plant was expensive, but still had a positive rate of return).

However, for CO_2, it is not going to work. First, there is the problem of transportation: (a) a single truck owner, driving 500 miles a day in a 5 GPM 18 wheeler, burns 100 gallons a day and produces about 1 ton of CO_2 per day or about 300 tons per year (there are about 26 million trucks); (b) a taxi driver with a 25 MPG car, driving 300 miles a day, produces 60 tons of CO_2 per year, and; (c) a typical drive to and from work produces 10 tons a year. Transportation accounts for about 1/3 of US emissions of CO_2. Not including transportation emissions defeat the efforts to eliminate emissions. The refiners (Exxon, Shell, etc.) would hire lawyers to argue that they are not responsible for the customers burning their products (gasoline or diesel). Second, even restricting ourselves to heavy industry, we are talking about thousands of companies, many of them with different needs and approaches.

142

One steel mill might use natural gas, but another might have electric furnaces. It might be possible to control the larger units within an industry, but there are thousands of small outfits using some natural gas and electricity that would also require permits to operate and obviously funds to modernize their operation. One cement plant might need more natural gas than another because of different limestone compositions.

Implementing cap and trade would be opening a can of worms. Limiting cap and trade to refineries, electric utilities and gas distribution companies might be easier, but they are only intermediaries, and as I am aware, they cannot reformulate their gasoline or natural gas to emit less CO_2; the electric utilities cannot afford to carry needed investments unless they substantially increase their rates starting today. If the regulators authorize the rate increases, the cap and trade permits are the excuse - the dreadful carbon tax - for them.

The issue of fairness is important. If we are giving, free of charge, emission permits to existing emitters, we are giving them an advantage that newcomers will not have. A solar developer needing natural gas for backup purposes during cloudy days would have to buy the permits on the open market, while the utilities that did nothing to develop solar energy get a free pass. Furthermore, the price increases of fuel and electricity would disproportionally affect the poor because energy is a bigger portion of their expenses.

Carbon Tax

This alternative has some advantages: (a) the government collects the taxes and therefore it can generate income to finance the investment needed to become a carbonless society, or subsidize renewable energies and even modify(!) the tax code to make the taxes fairer; (b) it could be imposed in the typical energy units discussed in the previous chapter and therefore people would pay the tax for using fossil fuels, which would eliminate wasteful consumption, incentivize reductions and improve efficiency; (c) it could be implemented across the board without major complications (gas stations and utilities already collect some tax, so it could be implemented quickly), and; (d) it would be perceived to be fair because it would be paid proportionally to the consumption of energy and those who make an effort to consume less energy would pay less.

It would cause high inflation because everything will increase in price, food, clothing, automobiles, etc., and people are going to demand higher salaries. Politically, it is probably the least palatable alternative. The Taxpayer Protection pledge was signed by 95% of all Republican members of Congress and all but one of the Republican presidential candidates running for the 2012 (and 2016!) election. In an ideal world, intelligent discussions would take place and a compromise could be reached. Given our current political climate, however, passage of appropriate legislation might take forever or never be implemented.

One way of making the carbon tax less disruptive and have a chance of being implemented is to rewrite the tax code and also tax energy consumption. Proponents might aim at a combination of carbon tax and income tax that should be neutral, somehow avoiding the problem of high inflation. But, if the energy tax rate is imposed to quench the demand for fossil fuels, the percentage of taxes applied in the future to renewable energy would increase, which then might need to be countered with a reduction in income tax to maintain a steady level of taxation.

Unfortunately, there is a need for some $600-700 billion a year to finance the investment. Rewriting the tax code might be good, but it cannot be revenue neutral if funds are to be collected. It needs to be a carbon tax to discourage the use of fossil fuels, and the carbon tax needs to increase over time, as less and less fossil fuels are needed. If it is to be simplified into an energy tax, the incentive to decrease fossil fuels disappears.

Rationing

The final alternative is rationing. Rationing is a nasty word that reminds people of past suffering in difficult times, for example, during wars. England had to issue rationing coupons for many items like tea, sugar and even bacon during World War II and even in the US there was rationing of tires and gasoline. However, it is the best way of distributing scarce resources evenly and fairly among people, when the free market economy would otherwise force the less fortunate to have to do without.

To reduce CO_2 emissions, the government could create a carbon emissions bank (the Emissions Bank – EB) and create linked accounts for every living person, issuing an emissions card (the EB card) to everyone. The EB would deposit monthly the prevailing amount of emission carbon rights into each person's Free Emission Rights ("FERI") account. Every person would be entitled to the same amount of emission rights. Parents of minors would be authorized to move funds from one account to another.

To be able to collect proceeds for the needed investment to become a carbonless society, to use the emission rights, people would have to pay the government a fixed price for each emission right. Upon payment, the EB would move the paid portion of the FERIs into the Paid Emission Right ("PERI") account.

The FERIs and PERIs would each represent the right to emit some CO_2 (the "Allocation Right"), which may be expressed as 1 or 60 kg CO_2 (close to the daily emission per person). The Allocation Right could be traded in multiples or fractions, just like dollars and cents. The two accounts (FERI and PERI) plus a regular checking account could be linked together the same way a checking account, an overdraft facility and a saving account are linked today. For those without banking access, the transactions could be done in post offices, grocery stores or at gas stations.

Those who need or want to consume more energy or emit more CO_2 would have to buy Allocation Rights from those who were more frugal or had smaller needs because they had modified their lives to use less energy and emit less CO_2. Soon, a market would develop for people to buy/sell PERIs. There would even be a futures market for PERIs. The price of the PERIs would be set by the market and could be displayed along with the Dow Jones and the NASDAQ indexes or in places where they could be bought or sold. Transport companies needing carbon rights could buy them in the market. Eventually, there could be businesses that would act as an intermediary for transactions, lending money to pay the government for the FERIs and automatically receiving the PERIs.

It is a more complicated system (more details will be given in the next chapter dedicated to the rationing mechanism), but it is fairer. Poor people would be able to sell carbon rights because they usually consume less. They could sell PERIs to make their house more efficient or exchange their gas guzzler for something that will consume less gas. Airlines would not have to buy PERIs. Their passengers would pay PERIs instead. For example, the airfare from Washington to San Diego could be $400 and 25 PERIs. A taxi driver would collect the fare and some PERIs (depending on the gas mileage of their car, which could be displayed along with the prevailing rates) and the customer could pay either all cash (with the PERIs paid for based on the current price or a defined proxy) or the cash portion with a credit card and the PERIs with their EB card. Transportation companies would tell their customers that their service would cost so many dollars and so many PERIs. If Coca Cola needed to distribute their product, they could buy PERIs in the market. The price set by the market would fluctuate with supply and demand. Strategies to hold selling allocations towards the end of the month might be popular and some hedgers would start figuring out ways to profit from it.

There are some risks. Identity thieves might get allocations also. For a small fee, a new insurance company might protect you and straighten up any mess. If people started hoarding their allocations, the price of the allocations would follow supply and demand and the price of the allocations would go up, eventually tempting the hoarders to sell. If Coca Cola were to have problems securing enough allocations, they might have to think of other ways of distributing their product in a more efficient way. Some companies might decide to fraudulently charge more allocations for their product and speculate in the allocation market.

Some industries might have to modify their pricing mechanisms or accept exposure to the risk of fluctuating allocation prices. Some products would have the price of the allocation embedded, which translates in a higher price. Others might decide to sell at a given price plus so much for allocations. With some products (produce or food in a restaurant), the only choice is to embed the allocation in the price.

The level of allocations granted every year would decrease over time, achieving the desired level of reduction of emissions. The market would be setting the price for the allocations and also deciding which companies should continue selling their products and which companies are producing products that are too expensive. If a can of Pepsi now costs $1.50 because of the cost of aluminum and distribution of colored water, maybe tap water is better for society. People would decide whether they want to buy additional allocations in the market or drive less, drive a smaller car, car pool or peddle a bike and sell some allocations.

It would be like creating a new currency to simultaneously operate in the market, with the PERIs circulating freely in the market, with an exchange rate defined by the spot price in the market, just like the price of shares, Euros or the spot price of crude oil. Everyone would be entitled to buy, sell or pay with PERIs. Most of the transactions could be done with an EB card (similar to a credit or debit card).

The system is extremely fair. Every individual would be allocated the same amount, regardless of color, sex, age, race, political affiliation, etc. Wealthier people who use more energy, with their several cars and large houses, would have to buy PERIs. Frugal, cost-conscious environmentalists or poorer people could sell their excess PERIs.

The system could be implemented quickly in developed countries which are the larger emitters. Poor people in poor countries might not have the banking access or level of sophistication needed to implement a rationing mechanism. But, it is also true that their per capita emission is small and they do not need rationing to decrease their consumption. What they need is financing to have access to renewable sources of energy.

Conclusion

I did it again. In a mere ten pages, I eliminated voluntary contributions, government regulations, the cap and trade mechanism and even the carbon tax from the alternative paths to a carbonless society, proposing instead a very complicated rationing mechanism that would result in a disguised carbon tax, with the incentive that a market would be created where some people would be able to sell some of their excess allocation rights to offset the cost of those allocation rights they have to use.

Since it is a complicated mechanism, the next chapter will be dedicated to details of the mechanism and its implementation.

Chapter 10 Carbon Rationing Mechanism

Objective: to have a fair way to control the emission of CO_2

Principle: Considering the dependence on fossil fuels in our daily lives, the fairest way to reduce emissions is by evenly rationing the right to emit CO_2. The mechanism would allow every individual residing in the US to emit a progressively declining amount of CO_2.

The mechanism would be administered by a newly created Carbon Emissions Bank ("EB").

Every individual, irrespective of age, sex, race, economic condition, religion or legal status would be entitled to receive equal size emission rights. Special provisions might need to be implemented for visitors (tourists), incarcerated individuals or persons in sanatoriums. Hospitals and funeral houses would have to report to the EB the name (and close relatives) of the new born and dead, respectively. The original list could be secured from the last census or from the IRS.

The EB would create two or more accounts for all individuals and issue to each one a card, the EB Card. The EB Card would be similar to a credit/debit card. Each card would automatically be linked to two accounts. The first one, the Free Emission Rights account ("FERI"), would be very similar to a savings account. The EB would deposit to each FERI account, free of charge, the current number of Allocation Rights Units ("ARU"), on the date that is assigned based on the first letter of the last name. The second account would hold the Paid Emission Rights ("PERI"). The cards of several family members could be linked together, with parents allowed to administer the accounts of their young children.

Each individual would be entitled to transfer Allocation Rights Units from the FERI account to accounts linked to it by paying the EB the prevailing tariff for each Allocation Right Unit. I will use the term ARU to refer to both FERIs and PERIs, since the

149

distinction between Free and Paid refers only as to whether the ARUs are freely given or have been paid for. The individual could pay for all or a portion of the FERIs in their account. The EB would then release the number (or fraction) of ARUs paid for from the FERI account to the second account, the PERI account. Payment for the FERIs could be made at banks, post offices, gas stations, at some grocery stores, at ATMs or online by transferring money to the EB. This is similar to sending an instruction to the bank asking them to move funds from the savings account to the checking account, or buying Euros from your saving account and putting them in a Euro denominated account (yes, PNC Bank offers the possibility of having an Euro account for business).

The PERIs would be fully transferable by using either the card or checks denominated in PERIs (which EB would provide to anyone requesting them), which could be deposited, electronically transferred or, using the EB Card, used to pay the necessary PERIs at a gas station or to the utilities providing gas or electricity or other uses or services that require an emission right (e.g. for airline tickets). The utilities and gas stations would use those PERIs to pay the suppliers of fossil fuels for the emission of the fossil fuels. The final suppliers of fossil fuels would have to balance their product sales with the PERIs received, justifying their sales with the collected PERIs. The PERIs collected by the fossil fuel suppliers would be extinguished (i.e. the fossil fuel suppliers would return them to the EB).

The PERIs can also be sold on the open market in the same fashion Euros, pounds or shares are sold, at banks, post offices, some grocery stores, gas stations or other venues interested in profiting from buying and selling PERIs, where the instantaneous market price of the PERI would be displayed prominently, just as the US dollar is quoted all around the world. The owner of PERIs would be able to walk into any of these venues and use checks denominated in PERIs at the current market rate or make an electronic transfer using their card. Eventually, somebody would offer to provide the cash needed to convert FERIs into

PERIs and then purchase the PERIs. A futures market would be created for PERIs, where large industries could buy contracts for millions of ARUs in the form of PERIs.

Any company needing to purchase fossil fuels for their processes or needs would create a demand for PERIs. Transportation companies, light and heavy industries requiring PERIs would create a market for them. Airlines and other transportation companies would charge rates broken down into dollars and PERIs, and might offer to sell customers PERIs needed at a price similar to the market price.

Statistically, 99% will have PERIs to sell, mainly because the ARUs take into consideration all demands for energy, but the typical family does not buy steel or aluminum or LDPE directly, but only indirectly when they buy a car, a house or an appliance.

There are complications and different needs, but rather than creating exceptions, it would be advisable that the price of their products reflect the added expense of the PERIs. So, rather than exempting taxis or farmers, it would be better to have their tariffs on products reflect the cost of the PERIs. Farmers need fuel for their equipment and fertilizers that are energy intensive and their products (corn, wheat, soybean, tomatoes, strawberries, apples or milk) would reflect the cost of the needed PERIs.

Requirements

The Emissions Bank would initially have some 320 million clients, each with at least two linked accounts, the FERI account and the PERI account. Many of those accounts could also be linked together into an account for the head of the household, so that accounts of minors could be managed by their parents. The EB would be responsible for issuing the EB Card to each individual, and therefore would have to have a database continuously updated with new births and deaths, plus the ability to link/unlink accounts as needed, so that when a minor reaches adulthood and chose to have their own card, the account can be unlinked from their parent's account, or, if a couple splits, their accounts could be unlinked at the request of either one of the parties, or marrying couples could join accounts.

The EB would issue monthly FERIs to all clients (about 15 million transactions per day); make millions of transfers daily from FERI accounts to PERI accounts (about 25-30 million transactions per day) between linked accounts; and receive millions of dollars a day. In all likelihood, the EB would be a major bank and might have to have offices everywhere. Points of sale, also linked to the EB might include post offices, banks, ATMs, gas stations, etc. The volume would be similar to that of Visa or Mastercard transactions. The administrative expenses might reach 1 to 2% of the collected amount.

Possible fraud includes requesting additional cards for non-existent family members, card theft, bounced checks, stolen identities, failure to report the death of a family member, disgruntled spouses, coerced transfers, counterfeited or falsified cards and others that imaginative thugs would find. Additionally, some fossil fuel companies might underreport sales or PERIs collected or some utilities or gas stations might request more PERIs than needed and sell them on the side. Illegal aliens, long-term visitors and even tourists who should also be subject to rationing, would be given FERIs and would have to pay the prevailing rate to convert them into PERIs. Just like a credit card at a gas station, the card reader could find out how much gasoline it can dispense; the same card reader could read the EC card and find out how much gasoline it can dispense.

Numbers

The numbers in this section are reasonable approximations. The US has 320 million people and emits 6.7 GTY of CO_2. The initial size of the ARUs should be for about 60 kg per person per day and should be reduced over time to reach zero at the end of the program. The allocation size is immaterial, but given the fact that we normally only use two digits after the decimal point, it would be advisable to have the ARUs denominated in small units, like the right to emit 1 kg of CO_2.

Using the emission of CO_2 per typical units discussed before, a family of four would be receiving some 7,200 FERIs per month (60 kg/day times 30 days per month times four members or the equivalent right to emit about 7.2 tons per month). If they used 1,200 kWh in the month, they would have to use 1,248 PERIs $(1,200 \times 1.04$ kgCO_2/kWh) to pay the electric company in addition to the electricity rate. If they filled two cars with 20 gallons of gasoline once a week, they would need 2,321.7 PERIs per month (20 x 2 x 4.2 x 13.82 kg CO_2/gallon), and that month they would need 3569.7 ARUs, but would have 3,630.3 PERIs to sell.

Let's assume that the government estimates that it would be possible to become a carbonless society in 40 years starting in 2020, and that the level of energy needed then will be equivalent to 80% of the demand in 2014, fixing the price of the carbon fee at \$90/ton of CO_2. Our family would have to pay \$321.27 to the EB to transfer 3.569.7 FERI to the PERI(s) account to be able to purchase gasoline or pay for the electricity, but would have 3,630.3 FERIs that they could transform to PERIs and then sell. If the spot price were 10¢/PERI (or \$100/ton of CO_2), they would have to disburse \$326.73 to pay the EB, but would be able to sell it in the open market at \$363.03. The end result for the family would be that they would have to pay the EB \$648 that month, but recover \$363.03 for a net cost of \$287.97 to maintain their standard of living. The less energy the family used, the more PERIs they could sell and therefore end up paying less. If the spot price was \$150/ton of CO_2, they would be able to sell their 3,630.3 PERIs for \$544.55, and their net cost (e.g. the \$648 paid minus the 544.55 collected) would be \$103.45.

The spot price cannot be below the set price of the carbon fee, otherwise nobody would pay the carbon fee to sell afterwards below the price they had paid. A few examples will illustrate how the industry would help create an efficient market for FERIs:

- A fertilizer plant needs 34 MBTU per ton of ammonia, requiring 34 times 67.8 kg CO_2/MBTU or about 2.3 tons of CO_2 per ton of ammonia. If the plant produces 2,000 tons/day, it will require 4.6 million PERIs (1 PERI = 1 kg CO_2) per day, which at a spot price of 10¢/kg CO_2, will cost $460,000, or about $230/ton of ammonia, which unfortunately will almost double the price of ammonia, and therefore the price of corn, wheat, soybean, apples and carrots will go up.
- A 747 plane requires about 30,000 gallons for a transpacific flight, at 13.8 kg of CO_2 per gallon, which means that it will produce some 414 tons, which requires some 414,000 PERIs for a one-way trip. Assuming 500 passengers, the ticket will cost $82.90 more to cover the value of the PERIs.
- An 18 wheeler will require some 600 gallons for a coast-to-coast trip, or about 6 tons of CO_2, or 6,000 PERIs, increasing the cost of shipping the freight by about $600. If the truck is carrying 100 TVs, the cost of the TV to account for the PERIs will have to increase by $6.

All business, large or small would need to buy PERIs because all of them require energy. A restaurant needs to run ovens, stoves, dishwashers, etc.; a grocery store needs lighting, refrigeration, etc.; a movie theater needs air conditioning and some lighting: etc. The initial amount of ARUs would be sufficient for everyone to acquire the PERIs needed for their direct and indirect needs. The initial amount should represent 100% of the previous year demand. Overtime, as we get accustomed and all the kinks are removed from the system, the amount of ARUs to be issued would be reduced.

Initial Use of Proceeds
The monies collected by the EB ought to be used to finance the transformation to a carbonless society. DoE would be in charge of setting up the work plan, assigning priorities, coordinating and

overseeing adherence to the plan. It would also be responsible for allocating the bulk of the monies to the states, proportionally to their contribution. The states would in turn allocate the monies to the cities or municipalities according to their contribution. Some money would need to be set aside for transmission. DoE would be responsible for overseeing the expansion of transmission of electricity and transferring to the states and/or municipalities the funds needed for the construction of transmission lines within the state/municipality.

It would be advisable to start with a small carbon fee to remove all the kinks from the system. Gas stations and utilities would be entitled to sell PERIs for those who have not received their cards or have not figured out how to make transfers or because the gas station has not yet installed the adequate system. Even setting the carbon fee at $1/ton, the system working properly would collect $6.7 billion in one year. The price to transfer all the monthly FERIs for our hypothetical family would be only $7.20 per month, more a nuisance than an actual cost. The initial funds would be needed to:

- Build the "banking" infrastructure and negotiate with Visa or Mastercard to accept the EB card issued by the EB, because 1% of hundreds of billions is a lot of money.
- Create an agency (the Renewable Energy Agency or REA) within the DoE, responsible for the program and hiring engineers and project managers to start planning. The project is similar to rural electrification or the interstate highway system, and would also require prioritizing the allocation of resources to critical points first (the effort is similar to gearing up the infrastructure to support the armed forces at the beginning of World War II).
- Create similar agencies at the state level (state REAs) to plan and coordinate the work in the states.
- Possibly build several factories all over the country to manufacture collectors, since huge areas would have to be covered by collectors, and therefore it might be advisable

to have a multitude of factories to reduce future transportation expenses.

- Similarly, establish local factories for about 2 million wind turbines and again reduce transportation costs.
- Create a budget for the program and propose a plan for ramping up the price of the FERIs to have the funds needed and coordinate with the state agencies the transfer of monies according to their budgets and plans.
- Help universities shift priorities to train many more nuclear, electrical, chemical, environmental, mechanical and civil engineers.
- Provide for drastic increase of research money in preparation for the program while the matter is being discussed in Congress/the Senate.

Obviously, the program would start slowly, reach a crescendo and then taper off slowly, just in time to start thinking about retrofitting some installations with newer, more efficient systems. The price to transform FERIs to PERIs should be changed as needed and as the amount of fossil fuels being burned decreases. If the amount of emitted CO_2 is reduced to 50% of the original emissions, the price of the ARUs would have to double to collect the same amount of money (assuming no inflation). If the amount of emissions drops to 10% of the original emissions, the price of the ARU would have to increase by 10 times. Maybe it could then be replaced by a metric for energy consumption, which probably makes sense because even when the specifications might have called for equipment to serve for 25 or more years, there would be degradation, plus new developments might require renovating the new renewable energy fleet at the end of its life. The investment would have to be financed by the new energy fee, similarly to the current charges for electricity or gasoline. The fact that it is renewable does not mean that the equipment will last forever.

Funding Research

To embark on such an ambitious program to become a carbonless society, we need a lot of fundamental and practical research, engineering work and creative minds, many of those minds belonging to entrepreneurs not stifled by working with conservative utilities or academia. It is a big challenge, and we should not embark on it with closed eyes or wishful thinking. In a typical project, the EPC (Engineering and Procurement Contract) cost varies in the range of 2-5%. If the program is expected to cost $30 trillion, that works out to between $600 billion to $1.5 trillion in engineering and procurement work. Just the number of engineers needed is mind boggling. We need many more engineers. Universities need to incentivize engineering work. Spending $100 billion or more is not a luxury. It is a necessity so that we have something robust, durable and reliable.

Cisco, Apple, Google invest about 15-20% of their yearly revenues on research. It is inconceivable, incongruous, unbelievable, shocking, ridiculous and perplexing to notice that the electric utilities and the DoE combined spend only about $100 million on research (0.02% of revenues), knowing, as they must know, that CO_2 contributes to global warming and there could be very negative consequences. Our politicians, the leaders of industry and the professionals working in the DoE and similar agencies have failed us by lacking the courage to stand up and prepare us. Many of them feel happy about their work and congratulate themselves for creating ARPA-E, the SunShot initiative and other programs with catchy names. The first RFP (Request for Proposals) of ARPA-E attracted more than 3,500 proposals. They only funded twelve.

SunShot aims for solar thermal energy at 6¢/kWh and would not fund research for a demonstration unit that could only offer 12¢/kWh. Solar thermal at 6¢/kWh hour is today a pie-in-the-sky dream. Let's say solar irradiation is 5 kWh/d and you capture 60% of it (the high end of current horizontal paraboloid sun tracking collectors) and use it in a thermal engine with 33.3% efficiency. You would be able to generate 1 kWh/m^2 d. Assuming 300 days a year and 25 years, it would generate 7,500 kWh per square meter. At 6¢/kWh, you could recover only $450 dollars in 25 years (with no money allocated for maintenance, interest or managing the assets). According to NREL, collectors cost about $390 per m^2 and thermal plants about $1,000/kW. By setting the bar so high, they discourage serious research (people with good and realistic ideas are shunned), but exaggerated claims or plain misrepresentations based on grossly optimistic assumptions might get funded, but fail to deliver. The DoE justified the goal, stating that the projects have to compete with natural gas plants, and since there is glut of natural gas due to cracking, the price of natural gas is low.

Research at private companies has always been more successful than research at universities or national laboratories. Private companies care about results, even if they do not understand the why. Academic research cares about the why and usually gets sidetracked. National laboratories and universities get grants for specific purposes and must write lengthy reports and cannot deviate following a discovered promising avenue if it was not previously described in the original proposal. Smaller companies are nimbler, can move faster and are usually the leading edge of research. I suggest that some of the funding should be given to venture capital to administer to provide needed flexibility.

The US has lost some of its competitive edge. For a while, we even had to import wind turbines, and now we have foreign firms building them here. PV panels are mostly made in China.

The electric sector sells about $410 billion a year. They could have easily dedicated 1or 2% of their revenues to research ($4 to $8 billion per year) when there were writings on the wall that many states were going to impose new renewable portfolio standards. The electric utilities have not prepared themselves for possible additional demand arising from electric cars, entrenching their positions and letting others take the risk. The latest PPAs signed have very little profit potential if everything goes well, but have stiff penalties if the system underperforms. The utilities keep insisting on PPAs for hundreds of MWs, believing that economies of scale play the same role with renewable sources, letting the prospects of genuine distributed generation take hold. Rather than taking an active role, leading research to solutions that might benefit them and the public, they take reactive positions to preserve their status and legal rights. As the program progresses, all utilities will become just distribution companies, renting wires, because they will not be generating any electricity.

No one can deny that there has been extraordinary progress, but the majority of it was financed by venture capital. The DoE and the national laboratories are staffed by the best and brightest scientists available, but somehow they have lost their perspective. It might be advisable to invest in several venture capital funds to administer some 50% of the funds, to attract entrepreneurs. DoE and universities are accustomed to work with programs that seek explanations, rather than results. The interest in developing multi-layered PVs to capture more irradiation is academically interesting,, but the problem of storage also needs to be solved, otherwise the PV panel, despite its main advantages (no moving parts), is only an academic curiosity.

There are many areas needing practical research, from smart appliances connected to a smart meter (a device that can sense the instantaneous price of electricity), able to postpone its demand until the price of energy is lower, to a stable grid with variable input, to large size batteries, recycling lithium batteries for home storage, hydrogen generation, storage and dispensing, photochemical and thermo-chemical reactions, sites for water pumping storage, sites for compressed air storage and demonstration plants for novel solar thermal approaches. There is also need for fundamental research for Generation IV nuclear reactors (fast breeder nuclear reactors to eliminate nuclear waste and capture 100 to 300 times more energy from uranium), nuclear waste disposal or elimination, or for improved efficiency in lighting, heating and appliances, elimination of parasite losses and even signals for eliminating non-critical loads in the event of shortfalls.

The challenge is how to maintain the standard of living we are accustomed to without fossil fuels. Transportation is a good example. We can reduce our standard of living and give up flowers from Colombia, Perrier water from France, fresh asparagus year round from South America and the bulk of the global trade, or we can find a way to make transportation much more efficient. A Panamax vessel consumes 21,000 gallons of diesel per day and a 747 consumes 16,000 gallons per transatlantic flight from New York to London. The weight of a hydrogen storage container would prevent planes from flying and take up a lot of cargo space on a ship. Both are highly optimized designs, so it is unlikely that they can be improved substantially. Are we going back to zeppelins and sailing ships, or do we invest in research? Maybe we need to expand the railroad service, or really accelerate dispensing hydrogen or find a way of having a system to exchange batteries along the interstates. If we do not enlist the help of thousands of creative people, we are not going to be able to come up with possible solutions.

The man on the Moon effort in 1964 is a good example. There were thousands who solved little specific problems, like designing fuel cells for electricity in space, a way to generate oxygen, CO_2 absorbers and propulsion systems and others who worked to provide faster communication, even if it meant putting stations around the globe. The need for small, light devices was the foundation of personal computers and digital cameras.

Chapter 11 Leading the World

We are not through yet with the complications and uncertainties. Even if we were successful in transforming the US into a carbonless society, if we failed to convince the world to follow our example, our sacrifices and expenditures would only postpone the inevitable. We need to convince and cajole the whole world to follow our example. Some countries would quickly join us in the efforts; some others would politely request a grace period until they reached a standard of living comparable to ours, and others would plainly refuse to join, to profit from the competitive advantage offered by continuing to burn fossil fuels or simply continue to sell oil, natural gas or coal.

The benefits of burning fossil fuels are obvious. The energy density is huge, the technology is mature and, in many senses, affordable. Even very poor countries have some trucks and other means of transportation, and have electricity in major population centers. Somehow, at times with the help of multi-nationals, they manage to bring to the world marketplace those products that are in worldwide demand, earning some foreign exchange to buy the oil they need. Poor countries are not going to be able to afford a switch to renewable energies at today's prices.

The Kyoto Protocol was agreed upon in 1997, almost 18 years ago, but the results leave a lot to be desired. The US never ratified the treaty, taking the wind out of the sails for others willing and ready to pay the high costs of implementing its modest goals. As a result, CO_2 emissions did not decline at all, but have continued their inexorable climb. Further meetings (there have been about a dozen subsequent meetings in exotic places) have failed to reach an agreement.

The negotiations are complicated. The developed countries ("D countries") want everyone to follow the agreed mandate; otherwise the less developed countries ("LD countries") have a competitive advantage and can capture more market share in the D

161

countries, resulting in greater unemployment for D countries. LD countries blame the developed world, with their prolific consumption of fossil fuels, of causing global warming (or the more polite term "climate change"), and suggest that the D countries first curtail their emissions and pay (or help pay for) the conversion to renewable energy of the LD countries. Putting it in a simplified form, the LD countries would like to set an emission level per capita of x tons per person year, with the D countries having to reduce their emissions from whatever they are emitting today to the level x. The LD countries can continue using fossil fuels until they reach the level x.

Since there are more LD countries than D countries, no agreement is reached. Some D countries with veto power at the UN and can veto any resolution they do not want.

Numerous white papers, position papers and negotiating strategies have been written and discussed here and there. I believe that the position of LD countries is sound and they have a point. The D countries defend their position, arguing that since nobody was aware of the consequences and there was no intent to harm, they cannot assume responsibility, and since we are now facing a common problem, we all have to work towards the solution.

The rationing mechanism might be a way to reach an agreement.

Worldwide rationing mechanism

I propose that we use a rationing mechanism at the world level. The basis is the same: a fair way to eliminate the emission of CO_2. Each person, regardless of country of origin or nationality, color of skin, age, religion, sex or political affiliation, is entitled to emit the same amount of CO_2 (the Allocation Amount or "AA"). Consequently, the number of people in each country defines the amount of CO_2 the country is entitled to emit (the Country Allocated Amount or CAA), by multiplying the population by the AA. Countries that emit more than the calculated CAA would need to buy AAs from countries that have excess AAs.

The implementation of a rationing mechanism in the US and other D countries is complicated but doable. The implementation in LD countries might not be possible because not everybody has access to banks, but it is not necessary. In LD countries, it could be implemented at the national level, since the LD countries are likely to emit less than the initial target, and therefore would be able to sell AAs to D countries emitting above their CAA.

The system is relatively easy to monitor. We have reasonable population census figures in most countries, and while they are not easy to update and verify, they are considered to be reasonably accurate. We also have statistics of population growth, and even if the numbers are not updated as often as desirable, projections should be able to produce reasonable approximations. The trade of primary energy sources is also followed by many organizations and data about shipments of crude oil, carbon and natural gas are available. Two other sources of primary energy do not produce CO_2. Nuclear energy is limited to a few countries and the hydroelectricity generation potential is known. The effect of burning wood or dung is not known, but thankfully represents only a small fraction of the world's energy budget. BP's annual report lists the primary energy use of more than 100 countries, the World Bank reports the emission per capita of most of their member countries and could ask for data for all their borrowers, and finally, the UN has a good handle on the population of each of its member countries.

Let's assume optimistically that the mechanism could start working in 2020. Each country has many years to produce the data needed to calculate their country's allocated amount, by carrying out a new census, updating population growth statistics and establishing mechanisms to report local production of primary sources and imports/exports of primary sources, therefore, being able to provide to the United Nations or to a newly created World Carbon Agency ("WCA") their estimated Country Carbon Emission ("CCE") data.

The AA can be defined by the total production of CO_2 in 2015, divided by the estimated population of the world also in 2015 or about 4.86 tons CO_2/p y. For practical purposes and to continue with the argument, let's use 5 tons of CO_2 per person per year. That means that Bolivia, which might have 11 million people in 2015, would be able to emit 55 million tons of CO_2 in the year 2020. Since, according to the World Bank report, Bolivia emitted 1.58 tons of CO_2 per person in 2010, which, roughly extrapolated to 2015, this means that Bolivia emitted 1.7 tons of CO_2 per person per year, producing a CCE of 18.7 million tons per year. Therefore, it would be entitled to sell 3.3 tons per person or 36.3 million tons of CO_2 emissions to the WCA. For the US, the picture is different. The US population in 2015 is rounded to 320 million and the emission per capita is rounded to 20 tons per person per day. The US would have a CCE of 6.4 billion tons, but with the CAA, it could only emit 5 x 320 million or 1.60 billion tons in 2020. The US would have to purchase from the WCA the difference between the CCE and its CAA by purchasing 4.8 billion tons of emission rights from the WCA.

World Carbon Agency

The WCA would validate population size and the emissions of all countries. It would require countries to implement systems that would allow easy verification of the data, carry sporadic audits and would be the final authority regarding the CCE and adjustments, if needed, to include changes in primary energy inventories. The WCA would collect monies due from countries needing additional AAs and distribute the proceeds among countries selling AAs. Having the WCA mediate the transactions removes extraneous conditions for the transfer of funds.

At least initially, the WCA would only disburse funds to the selling countries for renewable energy projects; however, it could graduate countries that have progressed faster on their renewable energy projects, and graduated countries could use the monies for other long-term infrastructure, health or educational purposes.

Furthermore, the funds might be only used to pay for imported components of renewable energy projects, unless the country's transparency index were above a threshold defined by the WCA, to ensure that the funds disbursed were effectively being used for the intended purposes. If necessary, the WCA would inspect and monitor project progress. This is similar to how the World Bank operates requiring that countries also contribute some funds for the projects, with the resources provided by the World Bank financing imported components.

The system should be easy to implement and monitor, which does not mean that it would be perfect. Some countries might exaggerate the size of their population to gain more allowances or might want to hide some of the emissions. Some countries might drag their feet or refuse to pay the amounts due to the WCA. Most countries requiring payment of excess CCEs are D countries with sound legal systems, and while many (e.g. the US) would need congressional or parliamentary approval, once the agreement was ratified, the monies should come regularly. However, some countries, especially oil exporters with large emissions, might object to paying for the excess AAs, maintaining that it is the countries buying the oil that are the ones responsible for it, similar to the argument used by the tobacco industry that it was the smoker's choice. Probably the only tool for pressing for delinquent payments would be the imposition of import or export sanctions, with the WCA publishing a list of countries in default of their obligations and imposing a ban on importing or exporting items from that country, with importers and/or exporters having to certified that they have checked the list of eligible countries and that the sales/purchases from that country comply with the resolution. Eventually, a country importing goods from a rouge country would have to remit a portion of the payment to the WCA or countries exporting to rouge countries would have to insist for payment of the surcharge required by the WCA.

Most OPEC members are among the largest per capita emitters and would have to pay large CCEs. Unfortunately, the consequences of a worldwide effort to reduce/eliminate CO_2 emissions would have severe economic consequences for oil exporting countries. Many of them are dependent on the sales for most of their foreign exchange needs to import food and other goods not available in the country, to pay for many government functions and to subsidize a growing population. Since the alternative of desperate moves to capture more foreign exchange - an oil price war - would be detrimental to all, the implementation of a carbon rationing program is likely to strengthen the OPEC, giving them more control over their output, which might result in a steady price increase to cover their payment for excess emissions, and also to finance the installation of renewable energy sources and the diversification of their economies in the most strategic way. Many of them are likely to underreport their carbon emissions to avoid having to pay large CCEs. Ultimately, it is the importing country who would be paying the respective portion of the excess over the allocated amounts, creating a price differential for oil from different countries. More diversified economies (e.g. Russia) have lower emissions per capita than Qatar or Kuwait.

The WCA is likely to become a powerful institution, similar to the World Bank, with many professionals monitoring population data or imports of oil, coal or natural gas, relying on the local governments to provide the bulk of the data, but checking that the data are reliable. Furthermore, the WCA would update the excess or deficit of the CCE and once a year provide adjustment accounts. For those countries showing a shortfall of the allocated amounts owed, the payments would be added in the forthcoming year and those countries that overpaid would be given a credit for the next year.

Payments to the WAC should be made once a month by countries needing additional emission rights, but disbursements to those countries selling allocation rights would be tied to the budgets or construction schedules for renewable energy projects. It is likely that at some point the WAC would end up having huge amounts of funds committed but not disbursed.

166

Timing and country groups

The timing is a simple variable and therefore easy to define. It does not mean that it is going to easy to negotiate or implement, but hopefully is one of the simpler things to negotiate. Considering the possible repercussions of climate change, the costs and the complexities of the project, it is recommended to have it done in not more than 50 years. Considering that the negotiations and the creation of the institutions would also take some time, let's assume that the project starts in 2020, and that the reduction is linear. As an example, if the agreed allocation amount is 6 tons of CO_2 per person year, in 2021, the allowed emissions would be reduced to 5.88 t/p y, for 2022 to 5.76 t/p y, for 2030 to 4.8 t/p y, in 2040 to 3.6 t/ p y and finally achieve no emissions in 2070.

The timing should be acceptable to most of the LD countries, because the majority of them emit substantially less than that. Furthermore, it would be advisable to stagger the implementation with three groups of countries, with the first group starting in 2020, the second in 2030 and the third and last in 2040. Even if the third group starts in 2040 with a ceiling of 3.6 tons of CO_2 per person year, the third group emits less than that amount today. Details could be discussed after the country groups are defined.

There are several alternatives for creating the groups of countries. One possibility is to create the groups ordered by per capita CO_2 emissions. With some exceptions, the largest emitters are oil producing countries, with oil/natural gas representing almost all exports. In 2011, Qatar was the largest emitter with 43 tons of CO_2 per person year, followed by Trinidad and Tobago with 37.2 tons of CO_2 per person year and then Kuwait with 29.1 tons of CO_2 per person year. Other oil producing countries with more diversified economies (e.g the US, Russia, Venezuela and Saudi Arabia) are also large emitters. Saudi Arabia, the largest of that group, emits 18 tons of CO_2 per person year. Another possibility is to select the groups in order of GDP per capita, with the idea that the richer countries should start investing earlier to become carbonless. The criteria used for the breakdown is shown in Table No. 21[18].

Group	GDP per Capita ($)	Population (million)	Population (%)	CO_2 emitted (GTY)	Emissions (%)
Group A	>20,000	1,027	14.17	11.76	35.46
Group B	<20,000 and >4,500	2,560	35.60	16.45	49.61
Group C	<4,500	3,621	49.96	4,945	14.91

Table 21 Characteristics of Groups of Countries

The data used for GDP per capita was obtained from the World Bank for 2010 (there are figures for 2014 for about 90% of the countries, but China has not yet reported data for 2014). The first group, Group A, is composed of 43 countries that have a GDP per capita above $20,000 per year. It includes a total population of 1.027 billion people, representing 14% of the world's population in 2014. As a group, they emitted about 11.76 GTY (the figures were calculated by using data from the World Bank, using the population of 2014 and the emission per capita of the last year the data is available – 2011), which represented 35.46% of the total emissions. Group B contains 58 countries with a GDP per capita less than $20,000 but more than $4,500. Group C comprises the remaining 112 poorer countries.

Group A includes Monaco ($145,221), Liechtenstein, Luxembourg, Bermuda, Norway ($87,646), Switzerland, Qatar ($71,510), Denmark, Macao SAR, Sweden, Australia, Netherlands, the United States ($48,374), Ireland, Canada, Austria, Singapore, the Faeroe Islands, Finland, Belgium, Andorra, Japan, Germany ($41,726), Iceland, France, Kuwait, the United Kingdom, Italy, the United Arab Emirates, New Zealand, Hong Kong SAR ($32,550), Brunei Darussalam, Spain, Israel, Cyprus, Greece ($26,862), Puerto Rico, Aruba, Slovenia, Portugal, South Korea, Bahamas, Oman and Bahrain ($20,546).

Group B includes the Czech Republic ($19,764), Malta, Saudi Arabia ($19,327), Equatorial Guinea, the Slovak Republic, Bar-

---

[18] GDP, Population and Emissions data from the World Bank and UN

bados, Trinidad and Tobago, Estonia, Venezuela, Croatia, St. Kitts and Nevis, Antigua and Barbuda, Hungary, Chile, Poland, Libya, Lithuania, Uruguay, Argentina, Latvia, Brazil ($11,318), the Seychelles, the Russian Federation ($10,765), Turkey, Gabon, Kazakhstan, Palau, Mexico ($8,916), Lebanon, Malaysia, Suriname, Romania, Panama, Costa Rica, Mauritius, South Africa, Grenada, Maldives, St. Lucia, Dominica, Montenegro, Bulgaria, Botswana, St. Vincent and the Grenadines, Colombia, Azerbaijan, Belarus, Cuba, Iran, Serbia, the Dominican Republic, Namibia, Peru, Jamaica, Thailand, Ecuador, Belize and China ($4,527). Given China's large population and emissions, the inclusion of China in either Group B or C changes drastically the weight and statistics of the group to which China is assigned, and since it is likely that, when implemented, it is going to be based on newer GDP per capita figures, and with China growth, it will be included in Group B.

Group C includes: Macedonia ($4,475), Iraq, Turkmenistan, Bosnia and Herzegovina, Jordan, Algeria, Angola, Tunisia, Albania, Fiji, Tonga, Samoa, El Salvador, Cabo Verde, Kosovo, Tuvalu, Indonesia, Marshall Islands, Armenia, Paraguay, Ukraine, Vanuatu, Swaziland, the Republic of the Congo, Guatemala, Guyana, the Federated States of Micronesia, Morocco, Egypt ($2,804), Mongolia, Georgia, Sri Lanka, the West Bank and Gaza, Nigeria, Bhutan, the Philippines, Honduras, Bolivia, Moldova, South Sudan, Kiribati, Zambia, Nicaragua, Sudan, India ($1,417), Papua New Guinea, Uzbekistan, the Republic of Yemen, Djibouti, Vietnam, Ghana, Cote d'Ivoire, the Solomon Islands, Mauritania, Cameroon, Sao Tome and Principe, the Lao People's Democratic Republic (Laos), Lesotho, Pakistan ($1,025), Senegal, Kenya, Chad, the Kyrgyz Republic, Timor-Leste, Cambodia, Comoros, Bangladesh, Tajikistan, Zimbabwe, Tanzania, Benin, Mali, Haiti, Nepal, Burkina Faso, Gambia, Afghanistan, Uganda, Guinea-Bissau, Rwanda, Togo, the Central African Republic, Sierra Leone, Guinea, Mozambique, Madagascar, Eritrea, Niger, Malawi, Ethiopia ($344), the Democratic Republic of the Congo, Liberia and Burundi ($220).

There are a few countries without data that were included in Group C. Those countries are: American Samoa, the Channel Islands, Curacao, the Cayman Islands, Greenland, Guam, the Isle of Man, St. Martin (French part), Myanmar, the Northern Mariana Islands, New Caledonia, the Democratic People's Republic of Korea (North Korea), French Polynesia, San Marino, Somalia, Sint Maarten (Dutch part), the Syrian Arab Republic, the Turks and Caicos Islands, Taiwan and the US Virgin Islands.

The grouping of countries is likely to be subject to difficult negotiations and will ultimately be decided with more current data and some adjustments. Many of the countries without data today would have to be moved to their corresponding group. Data from other sources could have filled the gap, but might be inconsistent. Some small countries in either group might have to be assimilated into other groups (e.g. Monaco, Liechtenstein, the Cayman Islands, Sint Maarten or the US Virgin Islands), but according to my philosophy, those are details or rounding errors that should not deter us from having a thorough discussion of the carbon rationing mechanism as a useful tool for curbing emissions of CO_2.

Allocated Amount

Before, as an example, the figure of 6 tons of CO_2 per person year was used. The best allocated amount to use during the negotiations is China's latest emission figure. The weight of China tilts any other logical limit. As discussed before, the emission figure per country was based on the last available figure of emission per capita, multiplied by the population in 2014. With total emissions close to 36 GTY, the halfway point would be 18 GTY. Prior to China's inclusion in the ordered list based on emissions of CO_2 per capita, cumulative emissions are 15.2 GTY, and with China, they become 24.4 GTY.

By placing the allocated amount at China's level, China would not have to pay to acquire allocated amounts from other countries in the first year, but would have to curtail its emissions at the 2% rate p.a. (the allocated amount reduced linearly over the period of

50 years) starting in the first year, otherwise, it would be required to start acquiring allocated amounts.

The size of the allocated amount affects the allowed CAA, and if the number is large, high emitting countries would have to buy less allocation rights, but LD countries would have more to sell. As an example, with 5 tons CO_2/p y, the US would need 15 tons CO_2/p y and Bolivia could sell 3.3 tons CO_2/p y. If the allocated amount were 7 tons CO_2/p y, the US would have to buy only 13 tons CO_2/p y, but Bolivia would have 5.3 tons CO_2/p y to sell. The initial allocated amount size is important because it defines which LD countries might need to start buying rights. I reiterate that it makes sense to put it at China's latest emissions.

Allocated Amount Price

The price of the AA is a transfer from high emitting countries to lower emitting countries which happen to be poorer countries that will need help to switch from their small fossil fuel consumption to renewable energy sources, hopefully at the same time raising their use of energy and standard of living.

In the Cost Estimates chapter, the estimated cost of switching all the energy required in the US to renewable sources and hydrogen for transportation was estimated at $30 trillion dollars. It was pointed out that one easy way of reducing the total cost would be to reduce the level of energy consumed. For the purpose of estimating the cost of the worldwide conversion, extrapolating the number from the US based on half the total emissions provides the ceiling of the cost, about $75 trillion dollars. D countries (and those emitting over and above the allocated amount) would have to finance their own conversion by implementing similar rationing mechanisms or a plain carbon tax. The LD countries with excess allocated amounts could finance the conversion with the proceeds from the sale of excess allocated amounts.

The AA price should reflect the urgency of converting mankind's need for energy to renewable, sustainable forms that would arrest the threat of climate change and its dire consequences, while at the same time being realistic and acceptable to the whole world. An ever increasing price would provide a strong incentive, both for D countries needing to purchase additional allocation amounts because the cost of the allocation amounts will be increasing and for LD countries because speeding the conversion would leave them more allocated amounts to sell at a higher price.

As an example, the proposed allocation price could start very low at $1/ton of CO_2 in 2020, but steadily increase $1 per year, reaching $51/ton of CO_2 in 2070. Other allocation prices and adjustment rates could be used and are likely going to be the result of difficult negotiations, with LD countries pushing for higher allocation prices and faster adjustment and D countries pushing in the opposite direction. The table below shows the estimated impact for two countries, the US and Bolivia, used as an example:

	Example of Transferring Allocated Amounts									
			US				Bolivia			
	Allocated					Cost				Revenues
Year	Amount	Price	Demand	Buy	Population	($ million)	Demand	Sell	Population	($ million)
2020	7.00	1	20.00	13.00	320.00	4,160	1.360	5.640	11.56	65.20
2021	6.86	2	19.60	12.74	322.24	8,211	1.380	5.480	11.68	127.98
2022	6.72	3	19.20	12.48	324.50	12,149	1.400	5.320	11.79	188.22
2023	6.58	4	18.80	12.22	326.77	15,972	1.420	5.160	11.91	245.85
2024	6.44	5	18.40	11.96	329.05	19,677	1.440	5.000	12.03	300.76
2025	6.30	6	18.00	11.70	331.36	23,261	1.460	4.840	12.15	352.86
2030	5.60	11	16.00	10.40	343.12	39,253	1.560	4.040	12.77	567.53
2035	4.90	16	14.00	9.10	355.30	51,731	1.660	3.240	13.42	695.80
2040	4.20	21	12.00	7.80	367.91	60,263	1.705	2.495	14.11	739.06
2045	3.50	26	10.00	6.50	380.97	64,383	1.531	1.969	14.83	758.94
2050	2.80	31	8.00	5.20	394.49	63,592	1.357	1.443	15.58	696.96
2060	1.40	41	4.00	2.60	422.99	45,091	1.009	0.391	17.21	275.80
2065	0.70	46	2.00	1.30	438.00	26,193	0.835	-0.135	18.09	-112.51
2070	0.00	51	0.00	0.00	453.55	0	0.661	-0.661	19.01	-641.16
						2,188,311				21,860

Table 22 Example of Allocation Price

The table shows that for the allocation price used as an example, the US would end up paying approximately $2.2 trillion during the 50 year period, with the amount paid peaking in 2046 at $64.7 billion, while Bolivia would be receiving a net amount of about $22 billion, peaking at $761 million in 2043. Because Bolivia belongs to Group C, it does not have to start reducing its CO_2 emissions until 2040, but if Bolivia does not expedite its conver-

172

sion, there might be years when it would have to purchase allocation amounts from the WAC. The calculations for Bolivia are
slightly off because some new electricity would be provided by
renewable energy that has not been included in this calculation.

There are some people that would balk at the amount that the US
might have to pay, which peaks at $64.7 billion and averages
about $43.8 billion per year. Putting the number in perspective,
when the US imported 10 million barrels of oil and the price was
$100/barrel, the US paid $360 billion per year. The peak amount
represents 0.43% of the US GDP, and the total amount represents
about 7% of the total cost of converting the US into a carbonless
society. For Bolivia, the sale of allocated amounts represents on
average about 2% of its GDP. Poorer countries would do even
better because they consume less fossil fuel and therefore emit
less and their GDP is smaller. However, as huge as $22 billion
would be for Bolivia, it is insufficient for its needs. For Bolivia
to have in 50 years the same consumption of energy that the US
has today, the investment in renewable energy needs to be close
to $1 trillion.

The calculations were repeated, multiplying the price per ton of
CO_2 times 7 so that the total paid by the US over the 50 year period represents approximately $15 trillion (1 GDP), which results
in the US paying on average about $300 billion, or 2% of its
GDP. The amount peaks at $452 billion (3% of GDP), which is
about the same amount the US paid for imported oil in 2008. At
those prices per ton of CO_2, Bolivia would receive approximately
$164 billion, which represents more than 5 times its current GDP,
or put another way, it would be receiving from D countries, on
average, assistance equal to 10% of its GDP per year. The effect
on Bolivia would be staggering. An infusion of 10% of its GDP
would have a tremendous effect on its economy. While the level
of assistance would not allow Bolivia to achieve the same standard of living as the US, it could finance renewable energy
equivalent to 16% of the level consumed by the US, an improve-

ment over the prevailing conditions today, with Bolivia consuming only about 6.5% of what Americans do.

It is expected that certain countries would object to the calculations for emitted CO_2, that others would complain that they are emitting CO_2 to sell to D countries and their emissions should be reduced accordingly, while others would point out that they are using the energy for mining for D countries or to extract and refine oil, or that they should join Group C because the exchange rate is wrong or they are tied to the Euro but really they are LD countries. The D countries would object to having to pay so much money and would like to start with $0.1/ton of CO_2, and to raise it only to $2, claiming that $24 billion is a large amount of money (for 4 billion people?).

The numbers suggested for allocated amounts and the price suggested per ton of CO_2 are based on logical arguments and appear to make sense. The matter, hopefully, could be subject to a series of constructive discussions and negotiations to reach a consensus and embrace the challenge of saving the planet and us. I hope that this book inspires others in a constructive way and that future generations thank us for having the courage to rise to the challenge and embark in this journey.

Chapter 12 Rising to the Challenge

Like many others before me, I despair and do not know how to continue. I know that the scare tactics of pointing out that we might be near the peak oil event or that there is a possibility that we might lose Miami or New York City to rising oceans does not work. People discount that possibility as absurd. The US might become the world's largest exporter of oil and surely we can find a way of stopping the oceans from rising. I know that serious discussions also get sidetracked with overly optimistic statements or other pressing matters. A friend of mine, an engineering PhD candidate, when talking about fuel cells and platinum, seriously told me that some companies were working on plans to mine asteroids to get the needed platinum. I ran out of arguments and wished him luck. I also talked at length with a retired head of a large public utility who seemed to show genuine interest in an idea I proposed to him and promised to get back to me real soon. After some weeks, I politely again sent him my business card in case he had misplaced it. I am still waiting for his call.

We can hope, pray or wish that technology will rescue us or take charge of our future and take steps to ameliorate the possible threats of global warming. Technology has been hard at work for the last 40 years (for batteries it has been more than a century) and what we have today is the best that technology has to offer. I do not have great hopes for a grandiose breakthrough in nuclear fusion in the next decade or two. The tools that I have recommended using here are the best, state-of-the-art ones we have now. They are far from perfect. Renewable energies cannot work 8,000 hours a year and they cannot compete in price with fossil fuels. I have been forced to include nuclear energy in the mix in order to have base load plants to carry us through cloudy or windless days. We also need liquid fuels for transportation, and I had dare to recommend that we use hydrogen, knowing quite well that the technology is still immature and that there are a myriad of problems with it, from generation, storage and even dispensing it, being aware that fuel cells are OK for the space station (not very sensitive to cost), but are not yet ready or reasonably priced for the massive needs of transportation.

We can give up, but Americans do not give up. They rise to the challenge and despite poor odds, they persevere and accomplish what others might think was impossible. I am appealing to those Americans who believe that purchasing insurance, even if it cost plenty is better than praying for miracles. I am appealing to those Americans who agree that a sacrifice for future generations is needed, that it is not only the proper thing to do, but that we have an obligation to do so.

For those who believe that we will be immune to global warming or to peak oil because of the mighty dollar or the powerful military apparatus, I am sorry to be blunt, but we cannot buy better weather, or rain for parched ground or snow for the Sierra in California or non-existent oil, and the military cannot fight mosquitoes or other pests surviving because of a lack of frost or from new pests invading our land. We are all in it together. We all live on the only planet we have.

Finally, we have to lead. We have seen early manifestations of some of the consequences, and science tells us that the trend is there and that the repercussions of not ameliorating it are not very attractive. Norway and Germany might be more concerned and have taken steps to reduce their contributions, but they do not have the weight that the US does, and few are following their example. If the US were to declare its intentions, start working on its implementation and discuss the possibility of impossing some kind of import tax on those countries that do not take steps to curtail emissions to protect their industries from unfair competition, the whole world would pay attention.

Let me summarize the findings:

- We have the tools that we need to effect change. They are not perfect and not cheap. We can have a mix of high temperature solar thermal energy with salts, dispatchable low temperature solar thermal energy and nuclear energy

176

in the mix as base load plants, and PV panels and wind energy producing all the energy we need, including hydrogen for transportation, to continue on our merry way.

- The cost is astronomical - $30 trillion dollars - but we could finance it with a carbon fee that would result in costs of energy on a par with European prices. The price could come down substantially if we desisted from wasting energy on snobbish things (we do not need to drive a three ton SUVs to go buy milk, or import water from a glacier in New Zealand). Small cars can take us out for milk, and water is water.

- A carbon rationing mechanism is fair, progressive and provides the proper incentives to motivate everyone to reduce their need for fossil fuels.

- What we need is a leader, someone who can convince the majority of Americans that we must make sacrifices for our children and their children and their grand children and great grandchildren. Someone who is willing to take action tomorrow, or the day after tomorrow at the latest. We cannot wait and risk the future of the human race because it costs money. We all know that it is cheaper to buy insurance than suffer the consequences. Finally, in a democracy, the leader does not have to convince everyone, only the majority.

- If properly motivated, Americans are capable of rising to the challenge and to start working towards becoming a carbonless society to prevent the pain, hardships and misery that continuing burning fossil fuels is likely to bring to the human race.

We are all together in the same boat. The boat is large – a sphere some 6,000 km in radius – but we are many light years away from any other habitable planet. We need to keep our boat afloat. Period. There is no other alternative. Either we act or we perish. I do not think that this is the end of humanity. But it might be the end of civilization as we know it. Some would survive in small rural clusters, but it would be a tremendous setback.

It is amazing that the US which rose to the challenge and fought two global wars to maintain freedom, the same US that rose to the challenge to reach the Moon, the same US that is exporting democracy, education and respect for human dignity and for women's rights, the US that has generously helped in the past, is now paralyzed because of petty interests, embroiled in polarized political infighting that prevents honest discussions about such an important issue, and refuses to act and lead.

Let's assume for a second that those who are worried are wrong. I am proposing to buy insurance. Invest in renewable energy. Arrest the emission of CO_2. If by any chance, unicellular algae were to exponentially grow and consumed the excess CO_2, well, we paid a hefty insurance policy and in the process became energy independent anyway. But if we just pray for a miracle and it does not happen, we have no recourse.

Process

The last proposal made by the Obama administration (August 2015) to regulate carbon dioxide emissions from coal thermoelectric plants met strong opposition, and even before the rule was announced, many states announced plans to fight it, including some vows to take the administration to court over the new rules. On the other hand, July 2015 was the warmest month on record and the year is on a path to be the warmest year on record. It appears that more people are connecting the dots and considering whether the drought in California and the awful fires in the Northwest might be linked to climate change and whether something needs to be done.

I anticipate, full of hope, that sometime in the near future, there will be a serious, furious, intense but real discussion about climate change, better known as global warming, answering tough questions such as:

How much time do we have? How do we train and provide jobs for displaced coal miners? How do we produce the thousands of engineers, surveyors, etc., needed? Should we place more solar facilities in the Southwest or let each state decide? How much money should be invested in research? How can we minimize the cost of the program? How should the burden be spread? When do we start? How do we handle inflation and price increases? Do we compensate public utilities or purchase assets? How can we avoid litigation with public utilities?

It will not be easy, and at times might be messy. There are going to be so many groups marching on Washington that the police might have to limit them to Monday and Wednesday only and employers are going to implement flexible hours to compensate for delays on the roads. Coal miners, taxi drivers, pilots and stewardesses, farmers, lobbyists, you name it and they will be there, because everybody will be affected, and while many are for saving the planet, nobody wants their position, situation or job to be affected, claiming that their welfare will be destroyed.

This book, just like others before mine, might be dismissed as naïve, unrealistic, uninformed or the product of a dreamer on drugs. The most likely outcome is that it will be ignored. If, by any chance, someone takes it seriously and dares to propose it as an alternative in some political event, the book will get some publicity and attract strong opposition. I might be accused of purposely misleading my readers, that my academic credentials are non-existent, that my grades at college were so so, or that I am a green dreamer, prophesying forthcoming apocalypses, but really only seeking money and glory.

Those are the problems of democracy. We should endure them and expect that in the end, rational decisions for the betterment of all and the planet will eventually be made.

Big Numbers & Small Numbers

There are many that will validly object to the high cost of the program and claim that $30 - $35 trillion is a heck of a lot of money, equivalent to $100,000 per person. They might want to create committees and sub-committees at the state and federal level to study alternatives, keeping everything continuing to work until a consensus can be reached as to the best alternative so that Alaska does not waste money on solar energy that would not work in that state, delaying action until the next election.

There will be others who will point out that $30-$35 trillion over 50 years, is only $600-$700 billion a year, and that we spend or invest (?) that amount yearly on the military. They will also point out that when the US was importing 12-14 million barrels a day and the cost of oil was $100/barrel, we sent our friends and others almost $500 billion a year. Finally, they will like to point out that most of the work will be done in the US and that it will have a multiplying effect, creating jobs, new industries, new careers and businesses, not to mention a livable future.

The critical questions we need to answer for ourselves and for the well-being of all the American people are:

What level of energy is really needed? Should we aim at replacing our typical wasteful ways or should we strive to become more reasonable? Do we need to continue driving SUVs or will smaller cars do? Both Japan and Europe use less energy than the US and seem to be happy. Maybe we can aim for a smaller energy demand and save plenty of money. Those who want more energy can now buy their own systems! They can install PV panels on their roofs, or get mid-temperature systems for their communities and even generate their own hydrogen.

Does it mean the end of prosperity, of well-being, of going out and having a good time? It might be the end of wasteful living, but it does not necessarily mean that it is the end of good times. We can have good parties with local produce without missing Peregrino water or imported cheese. To the horror of the French and the Germans, respectively, it will be possible to produce Brie cheese in Texas and have Heineken beers produced locally. We might not be able to change our automobiles every four years, but we will be driving less, with more efficient cars and with simpler fuel cell or electric cars. We do not need closets full of clothes we do not wear. Walking to the grocery store and interacting with the neighbors could start reforming the communities.

Can we still have a good time with less money? Every time you look back, you had less money then and yet, you got a house, bought cars, sent your kids to college and even saved some money for vacations or retirement. Before, when gas was less than $1 a gallon, you barely scraped by. When gas was $4 dollars, you survived. Nowadays, we do not blink an eye at paying $60-80 a month for a cell phone or $70 for cable TV or $69 for internet. In those three items alone, a family might spend $2,000 a year, but they don't go marching on Washington asking to be exempted. Everybody prioritizes their expenses to the best of their abilities and manages to have a good time and buy the essentials with whatever resources they have or make.

Is this the end of the American way? Only about 1% of Americans can go happily and merrily on with their lives without having to worry too much about the price of gasoline. For them, this is not the end. For many others, such as the typical middle class family, some adjustments will be needed, but it will not preclude sending their kids to college, taking vacations, eating out or watching cable TV. For the rest, the American way is a legend, that if you work hard, you can make it. I know that I would not be able to keep up with the hard work they do, whether it is picking cotton or strawberries, taking care of lawns, cleaning houses, working in shops or standing on my feet for hours at a time cooking or doing many of the jobs that pay only minimum wage. Many of them even work two shifts to make ends meet. For them, the American way is almost un-American, living from pay check to pay check, praying that there are no emergencies like a new car battery or new shoes for Johnny, paying a high interest rate for a clunker, high rent for an old and dilapidated apartment. Besides, the American way is not about wasting resources or being able to fulfill every whim. It is about everyone having the opportunity for an education, where everyone can start a business, where there is respect for the law and private property, where you can express your opinion without risking jail, where you can move from Florida to New Hampshire or California or somewhere in the middle. The American way in the 1800s was different from what it is today, yet it had the same ingredients. Anybody who wanted land could go west. Today, anyone can use inventiveness and entrepreneurial spirit to start a business.

I am not trying to negate that changes will have to be made. Some things will have to change. It might be cheaper (and more efficient) to use public transportation, so traffic jams are likely to disappear (hurrah!). Living in the suburbs is going to become passé, unless those suburbs transform themselves into small, much more self-sufficient communities, not too different from older, small rural communities. Our diet might change to what is grown and available locally, and we might have to learn to enjoy what the season has to offer. Cokes and other sodas are either going to disappear or be available only as concentrate to avoid shipping water, even if locally bottled. We will stop finding products from China or Europe on the shelves. Globalization is going to suffer. We might not be able to go to Rome or Japan as easily. Air travel might be prohibitively expensive. But there will be a resurgence of new jobs and opportunities for the American people, installing collectors, making furniture, clothes or shoes, inventing new plastics from ethanol, repairing things rather than buying new ones.

Side Effects

There is going to be a plethora of side effects, some good and some not so good.

There is going to be a substantial stimulus package, of unprecedented long-term duration, similar to the buildup of rural electrification or the interstate highways, and there will be plenty of opportunities for engineers, surveyors, construction workers, in the transport, manufacturing and electronics sectors, and for companies. A flow of $600-700 billion per year into the economy will provide jobs for millions, with most of the jobs at home because construction will be local, and the local industry will seek to capture most of the equipment manufacturing. In large projects and in manufacturing, labor accounts for about 1/3 of the cost, therefore it is likely that more than 50% of the investment is going to be distributed as salaries.

There is going to be some normal inflation because the added money will increase the demand for some goods; there is also going to be some forced inflation because the carbon fees are going to affect the prices of everything, from produce and crops to transportation, minerals and everything that needs energy to be produced; there are going to be price increases due to shortages of some commodities or specialized raw materials, but also due to some speculation by savvy traders. The inflation will put pressure on salaries, which will climb because there will be high demand for workers.

There are going to be losers and winners, with some coal miners having to learn new trades and contractors getting more jobs than they can handle, having to hire and train new people, and entrepreneurs finding way to reduce the cost of this item or that installation. There will be a reduction in demand for taxis that will become more expensive, which will reduce the number of taxi drivers but public transportation will expand significantly, requiring more drivers and buying more buses. Income from toll booths on roads and bridges will be reduced and the price might have to go up. Demand for airline travel will shrink, forcing many resort facilities to offer attractive deals to get people in.

The lawyers are going to have a great time. Demands from public utilities will be gruesome. Local neighborhoods will want to keep projects away from their back yard or protect their "pet" fauna. Even politicians will have a field day, promising to oppose all the negative effects, but keep the positive, or bring home the bacon by highlighting the strong winds or sunshine in the district. Universities and research centers will be fighting for research dollars. Companies are going to announce magically clever items that capture wind, light or heat more efficiently. This is the free market at its best.

Worldwide, it might be the catalyst that unites us. We are together in the same boat, and nobody will accept that some rogue states do not abide by the common goal of eliminating fossil fuel burning, irrespective of political inclinations, religious beliefs or economic situation. Oil producing countries will notice the drop in demand and might try to prop up the price of oil to gather funds to reshape their economies, which might result in a price war that is not good for them. However, they can eliminate their emissions quickly and be entitled to sell allocation units, and adhering to the program, they might benefit more than by selling oil to rogue states. The rogue states can adhere or be blacklisted by the world, embargoing them until a clear bill of health is given by the World Carbon Agency. The discussions as to the price of the Allocation Units, membership in the different groups and transfers from the buyers to sellers of allocation units, under the supervision of the WCA, of the use of funds for their intended purposes, will result in poorer countries receiving much needed help to install energy generation and distribution systems and hopefully enjoy energy that they did not have before. The example of the transfer of funds between the US and Bolivia showed that Bolivia could receive some $21 billion, which is roughly 2/3 of their current GDP, but if the negotiated price for the ton of CO_2 were to be such that the US would spend its GDP spread out over 50 years, the funds to Bolivia would represent more than 5 times their GDP, encouraging them to quickly stop emitting CO_2. Poor countries will sense they are important to the human race and will gladly contribute to alleviating the threat of global warming. The cost to the US and other developed countries is extremely high, but they must assume responsibility for the bulk of the emissions.

The transfer of funds will alleviate tensions worldwide and hopefully bring peace. It might sound that I have turned into an idealistic or utopian dreamer, but hope is the last thing to die. We probably will not have to fear the possibility of war over a diminishing oil supply. No disbursements will be made to countries if the conditions do not guarantee that projects will be completed. The transparency requirements of the WCA to disburse funds to the countries will spread democracy (dreaming again?). The WCA will not tolerate corruption. The funds are earmarked for renewable energy projects and not to enrich a few while the country continues emitting CO_2. Maybe we can achieve Utopia, having a sustainable, peaceful world, with poorer countries lifted out of poverty with the cash infusions from the CO_2 transfers, and we might even start saving money on defense.

Conclusion

Continuing to put our finger between our eyes and the Moon does not banish the Moon. It was demonstrated sometime in the eighteenth century that burning coal releases CO_2 and the conclusion that burning a lot of fossil fuels releases a lot of CO_2 is logical. The CO_2 concentration in the atmosphere has shown a steady increase, and the conclusion that we have overwhelmed the natural cycle is obvious. At the beginning of the twentieth century, it was demonstrated that CO_2 blocks infrared emissions, thus trapping heat. Some predictions of the effects of a warmer planet are starting to manifest. Whether you are a scientist or not, if you are intelligent and have an open mind, you can arrive to the same conclusion, that burning fossil fuels is warming the planet and that some of the effects will affect all of us negatively.

Humanity faces its biggest challenge ever. How can we maintain or even improve our standard of living without fossil fuels?

With some reasonable assumptions, I have concluded that the tools we have today could provide the energy that we need, that while the costs are high, that there is a fair mechanism to allow us to finance it and gain energy independence at a cost that will put our energy costs at Europe's level.

Obviously my work needs refinement. We may need more or fewer collectors or nuclear plants or wind turbines. The cost might be more or might be less; carbon rationing might or might not hamper the demand for energy. The first cost estimates of sending a man to the Moon were not very reliable, but we got there, and as a result, we developed a myriad of wondrous devices that have changed the way we communicate and interact with our families and with the whole world. We now have instant news from all corners of the world, in color on our TVs. It was just because we had the courage to leave the planet. Rather than looking at the negatives, try to imagine the benefits we can get for saving the planet. The GDP might stay same, but the happiness index is going to improve. We might not have globalization to bring cheap clothes from China, but we could have happier, smaller communities with satisfying jobs from craftsmen and women producing high quality items that last for generations.

I urge all of us to embrace the efforts of saving the planet for ALL mankind. It is time to face reality. Vote for those prepared to tackle and solve the problem. Let's get America leading the world again.

Note to politicians:

I started this chapter stating that I do not know how to continue. I do know. It is a lesson I learned early in my career as an engineer. I learned that the decisions are not technical, but economic, and I switched to the financial area and worked for the World Bank and for a private equity fund. I learned that the decisions are not only economic but also political. I do not have the character and or inclination to run for office.

I have tried for years to make a contribution to the development of renewable energy and to convince investors by appealing to: (i) their altruism to assist in leaving behind a better world for their children; (ii) their fear that peak oil is imminent or that global warming was going to result in less food, flooding of coastal cities and massive displacement of people, or: (iii) their greed, offering the potential of having attractive rates of return on sustainable projects. None of the approaches worked.

I am left again with the conclusion that the solution is political. Therefore, with all due respect and humility, I would like to remind you of your duty: You have been elected to represent all the people in assisting the government's role of fostering the welfare of ALL citizens, and while you might think that you were elected because your personal set of beliefs and philosophies on limiting the influence of the government in our daily lives or the level of taxation or the allocation of funds in the budget, the sad reality is that you might have been elected despite those beliefs, because the people trusted you more than they trusted your opponent. They might not agree with your views on abortion, gay marriage, gun control, Obamacare or your priorities in allocating funds for defense or education. They have elected you because they trusted that you would always vote for what you believe is best for them (all of your constituents), given the circumstances, even if at times it means increasing taxes, approving regulations or voting against your own party.

You have in front of you a situation that requires you to use your brain to seek the best solution to assure that we prevent some possible catastrophic consequences to mankind. Whether you are a Republican, Democrat or independent, you have an obligation to those who elected you to seek the best alternative. You have no obligation to lobbyists who are not elected, but paid for by others to try to influence you. They will always appear to you friendly and knowledgeable. They are trained to appear friendly to both Democrats and Republicans, congressional and senatorial representatives alike. They are knowledgeable because they have been groomed by the industry they represent. However, they do not have the best interests of ALL the American people at heart. They have only the interest of those that pay their fees.

You have an obligation to the welfare of your constituents, and that means that you need to rise to the challenge and work hard to assure that we, the United States of America, rise to the challenge, and your people and their descendants inherit a viable world with the same standard of living we enjoy or better.

www.ingramcontent.com/pod-product-compliance
Lightning Source LLC
Chambersburg PA
CBHW051907170526
45168CB00001B/285